Welcome

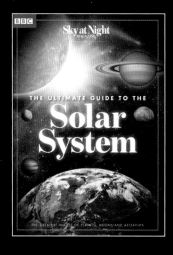

Thirty-five years ago, we had just set out to explore the farthest reaches of the Solar System. The Apollo missions had already landed on the Moon, while Mercury, Venus and Mars had been visited by space probes. Then, in 1979, Voyager 1 reached Jupiter on the first leg of a journey that would eventually take it out of our planetary system. Voyager 2, its sister spacecraft, would visit Uranus and Neptune. Over the following years, our knowledge of the planets orbiting our Sun would grow beyond all bounds.

During the subsequent decades following the success of the Voyager missions, many other probes explored the Solar System. Magellan and Venus Express mapped Venus. Galileo went to Jupiter. Cassini-Huygens set off for Saturn. Mars was visited by an armada of craft: Mars Pathfinder, Mars Global Surveyor, Mars Reconnaissance Orbiter, the Mars Exploration Rovers Spirit and Opportunity, Phoenix and, most recently, the Mars Science Laboratory, Curiosity.

Nor were smaller bodies neglected. NEAR Shoemaker landed on an asteroid, the Stardust mission collected cometary dust and returned it to Earth, and Deep Impact collided with a comet's nucleus. Meanwhile, an array of instruments studied the Sun. Throughout this time, the Hubble Space Telescope provided astronomers with a unique platform for detailed study of the planets.

As a result, scientists have presided over an explosion in our knowledge about the Solar System. The outcome is that their understanding of our home in the Galaxy is growing by the minute, and we now have a fuller picture of the bodies that orbit our Sun than ever before.

We've collected the most important of these images in this special edition of *BBC Sky at Night Magazine*. In the following pages you'll find stunning photographs from the spacecraft sent out to survey the planets, bringing you up to date with the current understanding of our cosmic back yard.

It's a mine of information and visual majesty like no other, and one which serves as a reminder that our home planet, Earth, is truly unique.

Chris Bramley
Editor

Editorial
Editor-in-Chief Paul McGuinness
Editor Chris Bramley
Subeditor Rebecca Candler
Editorial Assistant Emma Jolliffe
Writers Paul Bloomfield, Chris Bramley, Elizabeth Pearson

Art & Pictures
Designers Sam Freeman, Lisa White
Picture Editor James Cutmore
Picture Researcher Rhiannon Furbear-Williams

Press and Public Relations
Press Officer Carolyn Wray
0117 314 8812
carolyn.wray@immediate.co.uk

Production
Production Director Sarah Powell
Production Managers Louisa Molter, Rose Griffiths
Reprographics Tony Hunt, Chris Sutch

Circulation / Advertising
Circulation Manager Rob Brock
Advertising Director Caroline Herbert

Publishing
Publisher Andrew Davies
Publishing Director Andy Healy
Managing Director Andy Marshall
Chairman Stephen Alexander
Deputy Chairman Peter Phippen
CEO Tom Bureau

Like what you've read? Email us at bookazines@immediate.co.uk

Photo credit: NASA/JPL, Thinkstock

recycle
When you have finished with this magazine please recycle it.

IMMEDIATE MEDIA co

Contents

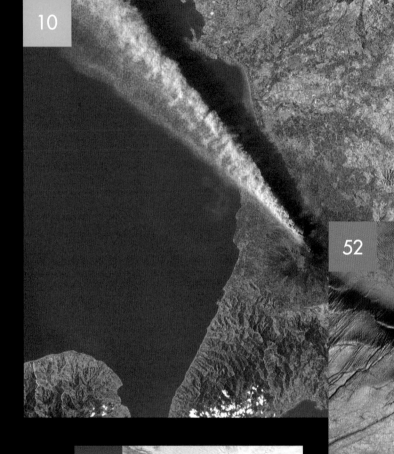

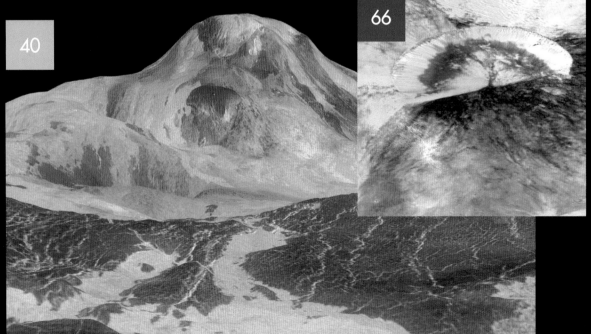

Contents

The Solar System

The array of planets, moons, asteroids and other objects orbiting our Sun ranges in size from specks to the 140,000km-diameter gas giant Jupiter, and stretches across over 18 billion kilometres of space

Photo: Thinkstock

Mercury

Venus

Moon

Earth

Mars

Jupiter

The history of the Solar System

F ive billion years ago, something was stirring out in space: a huge cloud of hydrogen and helium was collapsing. The gas rushed towards the centre of the mass, fusing together until it burst into life as the star that we now know as the Sun.

As the Sun was forming, so were the planets. Before our star was born, another, larger one had died in a supernova, filling the cloud with gas and dust. This debris gradually formed a protoplanetary disk – a huge, flat ring comprising hundreds of lumps of rock and ice known as planetesimals.

These planetesimals were the building blocks of the Solar System. After a few million years of crashing and melding together, these bodies began to resemble the planets as we know them today.

Close to the Sun, temperatures were too high for volatile chemicals such as water to remain solid in any quantities. The initial protoplanetary disk contained only a small amount of rocky solid material, so the four planets that formed closest to the Sun were comparatively small.

However, 750 million kilometres from the Earth at what is now the outer edge of the asteroid belt, temperatures were cool enough for gases to form thick atmospheres around rocky cores, creating the gas giants – Jupiter, Saturn, Uranus and Neptune.

It wasn't just planets forming, though; several moons did, too. Though many moons are former planetesimals that were captured by a planet, a few – including our own – had a much more violent beginning. When the infant Earth collided

with another young planet, a huge plume of debris was trailed behind. After a few hundred million years it melded together to create our planet's largest companion.

By four billion years ago the planets and moons had been formed, but the Solar System still looked very different from its current state. There were probably many more planets than the eight we know today, and they would have been much closer together. Over time the outer planets began to move slowly away from the Sun, throwing the gravitational forces of the Solar System off balance. The result was that several early planets were thrown out into deep space and, around four billion years ago, the remaining debris was pelted against the planets.

This period, now known as the Late Heavy Bombardment, left scars that can

still be seen on the faces of the Moon, Mars and other rocky planets. On Earth, such craters have been hidden by the actions of volcanism or worn away by the atmosphere.

The most significant relic left on our planet from that bombardment is the array of elements left behind. During Earth's formation, metals such as gold and copper sank to the core, so the deposits we find in the crust today must have arrived on asteroids and comets at a later date.

Perhaps the most important delivery to our planet was water. The early Solar System was far too hot for water to settle but, by the time of the Late Heavy Bombardment, temperatures had dropped significantly. When comets crashed into the surface of the early planets, water didn't boil off immediately but instead formed oceans.

"Several planets were thrown out into deep space and, four billion years ago, during a period now known as the Late Heavy Bombardment, debris was pelted against the planets"

After hundreds of millions of years, the planets had settled into their orbits and began to grow and evolve. Volcanism shaped their surfaces while, deep inside, molten cores began to cool. The cores of the smaller terrestrial planets solidified; without the flow of metallic cores, their protective magnetic fields faded, leaving their atmospheres unshielded from solar winds. As time progressed, such differences between each world became exaggerated, leading to the variation in planets that we see in the Solar System today.

And the process is far from over. Comets and asteroids still pelt the planets, and the Sun is slowly expanding and becoming brighter. In another few billion years the Solar System will have transformed itself once again. **Ⓢ**

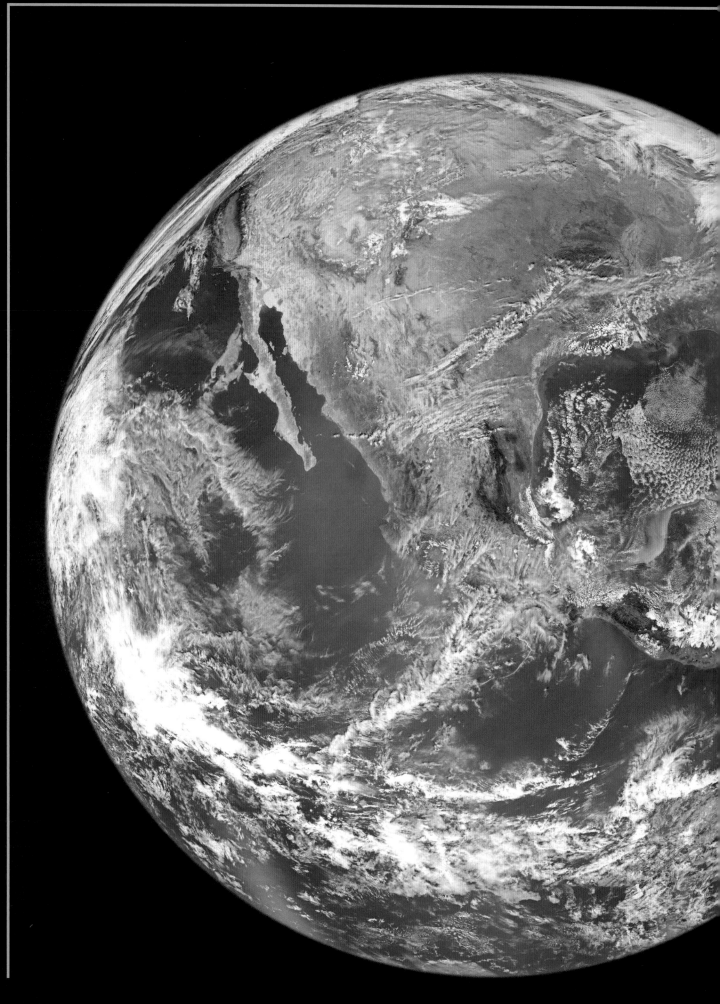

Earth and Moon

While the rest of the Solar System is populated by worlds of extremes, the Earth maintains a perfect balance. Not too close to the Sun to boil, not so far away that its water all freezes, it remains at just the right temperature to preserve the oceans that have helped shape our world. Indeed, our liquid water has been preserved for over four billion years, thanks to the atmosphere and magnetic field that protect us from harsh solar winds.

Perhaps our world's most striking feature is its Moon. Created by a colossal crash between a Mars-sized planetoid and the proto-Earth, ours is the largest natural satellite in the Solar System relative to its planet's size. The two bodies have a huge affect on each other. The Moon's pull on our oceans creates the tides, and our planet's pull has gradually reduced the Moon's momentum so that its rotation is locked – which is why we always see the same familiar face when we look up at the Man in the Moon. Ⓢ

Blue marble Earth

4 January 2012

A collection of photos taken by the Visible/Infrared Image Radiometer Suite (VIIRS) aboard the Suomi NPP satellite during four orbits of Earth was projected onto a globe and stitched together to create this picture. It is one of a number of full-disc views created by NASA and called Blue Marble Earth montages – images of mesmerising detail and beauty.

Photo: NASA

⊕ Cloud streets and von Karman vortices above the Arctic

24 February 2009

Cold northerly winds hit moist air over the Greenland Sea, creating parallel ranks of 'cloud streets'. The spiralling eddies formed as the wind diverges around the island of Jan Mayen are known as von Karman vortices. This image was captured by the Moderate Resolution Image Spectroradiometer on NASA's Aqua satellite.

Photo: NASA/Jeff Schmaltz/MODIS Rapid Response Team

⊕ Retreating sea ice

21 September 1979 (top)
16 September 2012 (bottom)

The north polar ice cap expands in winter and retreats in summer – but the retreat has become much more extreme over the past few decades. These two images, from sequences composed by NASA with the Goddard Space Flight Center, show how the minimum extent of sea ice has shrunk since 1979.

Photo: NASA/Goddard Space Flight Center Scientific Visualization Studio

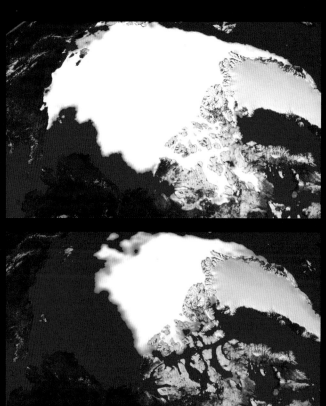

Hurricane Katrina strikes the Gulf Coast

28 August 2005

Katrina was one of the most powerful storms ever to hit the USA, with winds reaching 257km/h. This image was captured by the Moderate Resolution Imaging Spectroradiometer aboard NASA's Terra satellite as the storm approached the Gulf Coast, where it devastated parts of Mississippi and Louisiana, particularly New Orleans.

Photo: NASA/JeffSchmaltz/ MODIS Rapid Response Team/ GSFC

Etna erupts

22 July 2001

After six years of intense
activity at Etna's summit
craters, in July and August
2001 the volcano produced
a major flank eruption. Ash
streamed from a crater and
fissures on its southern slopes,
rising over 5km into the
atmosphere and casting a
dark shadow over the Sicilian
town of Catania. This photo,
with south at the top, was
taken by the Expedition 2
crew aboard the International
Space Station.

Photo: ISS002-E-8683 Space Station
Alpha/Earth Sciences and Image
Analysis Laboratory/Johnson
Space Center

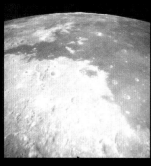

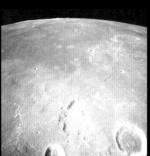

Mare Tranquillitatis

December 1972

The so-called 'Sea of Tranquillity' is a 'mare' – not water but a dark basaltic plain, one of several formed by ancient volcanic activity that cover about 16 per cent of the Moon's surface. These three images were taken by the mapping cameras on the command module *America* of the Apollo 17 mission – the last to land on the Moon.

Photos: NASA/James Stuby

Crater Tycho's peak

10 June 2011

Tycho is not the widest crater – though it's a not-insubstantial 82km in diameter – but is one of the most prominent, and contains this 15km-long massif, rising to 2km high. NASA's Lunar Reconnaissance Orbiter Camera captured this dramatic sunrise view of the mountains.

Photo: NASA Goddard/Arizona State University

Crater Clavius

With a diameter of about 225km, Clavius is one of the largest craters on the visible side of the Moon – so big it can be clearly seen with the naked eye. It's also among the oldest, formed some four billion years ago. This photo was captured by NASA's Lunar Reconnaissance Orbiter Camera aboard the robotic craft that has been orbiting the Moon since 2009.

Photo: NASA/GSFC/Arizona State University

The aurora borealis, as seen from the International Space Station

29 September 2011

If the lights of the US Midwest weren't enough of a spectacle, this photograph, taken by an Expedition 29 crew member on board the International Space Station, also captures the ethereal green glow of the aurora borealis as it lights up the night skies of Canada.

Photo: NASA

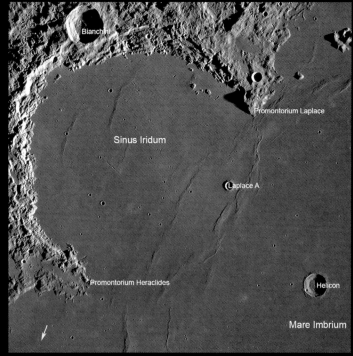

Apollo 15 mission on the Moon

1 August 1971

James B Irwin, pilot of the lunar module Falcon, salutes the US flag during the Apollo 15 mission. The fourth manned landing was in the Hadley–Apennine region at the eastern edge of Mare Imbrium.

Photo: NASA/David R.Scott

The Straight Wall

Rupes Recta, commonly known as the Straight Wall, is a linear fault or rille that forms a 300m-high escarpment stretching for 110km. This image was taken by the Reconnaissance Orbiter Camera.

Photo: NASA/GSFC/Arizona State University

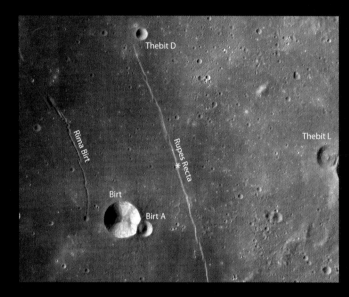

Sinus Iridum

The 'Bay of Rainbows' is a mare (basaltic plain) with a diameter of 236km, the remnants of a vast impact crater and itself pocked with other craters. The Sinus Iridum was the site of the first lunar landing since 1976: the Yutu ('Jade Rabbit') rover from China's Chang'e mission landed here in December 2013. This Lunar Reconnaissance Orbiter image is marked to highlight the major topographical points.

Photo: NASA/GSFC/Arizona State University

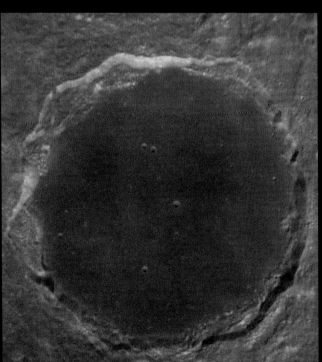

Crater Plato

Another big crater, Plato is 109km in diameter; filled with a dark basaltic mare and lined with a bright, irregular rim, it stands out clearly on the Moon's surface and has long been of interest to astronomers. This photo was taken by the Lunar Reconnaissance Orbiter Camera.

Photo: NASA/GSFC/Arizona State University

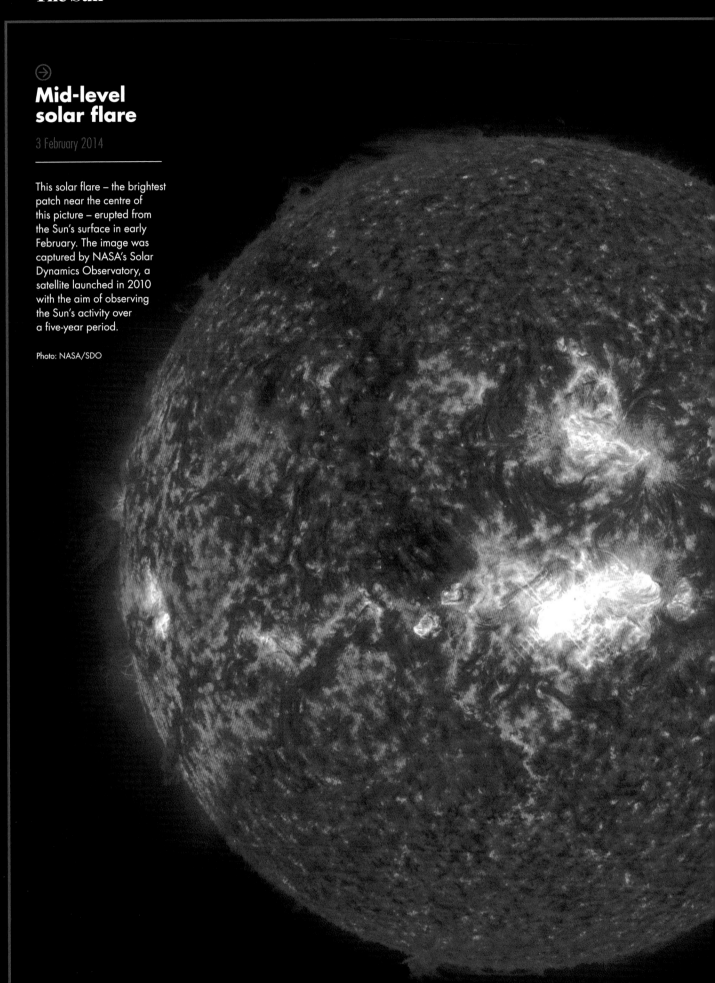

Mid-level solar flare

3 February 2014

This solar flare – the brightest patch near the centre of this picture – erupted from the Sun's surface in early February. The image was captured by NASA's Solar Dynamics Observatory, a satellite launched in 2010 with the aim of observing the Sun's activity over a five-year period.

Photo: NASA/SDO

The Sun

A t about 1,400,000km in diameter, the Sun is the biggest and most influential body in its system.

Weighing some two million trillion trillion kilograms, it makes up 99.98 per cent of the mass in the Solar System, though in astronomical terms it is only a medium-sized star. This huge bulk causes intense pressure in the Sun's core, where temperatures can reach over 1,000,000°C. By the time the plasma (charged hydrogen and helium particles) reaches the surface, however, it has cooled to a mere 5,500°C – still white hot, emitting light powerful enough to provide the Solar System with its energy.

The Sun's extreme conditions create a highly dynamic object. Its outer layers are constantly being shed, creating a solar wind of high-energy particles that sweeps through the Solar System. Meanwhile, inside the star, the plasma's motion creates a giant magnetic field that extends 18 billion kilometres from the Sun. Every 11 years this field flips, causing matter from deep inside the star to bubble up to the surface, exploding in spectacular solar flares. Ⓢ

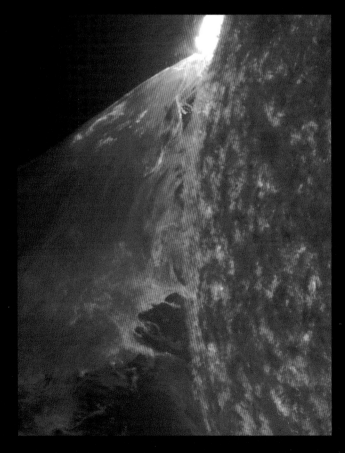

Solar prominence

24 February 2011

This large flare blew out a billowing flume of plasma that writhed for some 90 minutes in the corona – the Sun's hot outer atmosphere. Large, stable prominences can linger in the star's corona for several months and reach out hundreds of thousands of kilometres, sometimes draining back into the surface. NASA's Solar Dynamics Observatory captured this very high-definition image in extreme ultraviolet light.

Photo: NASA/SDO

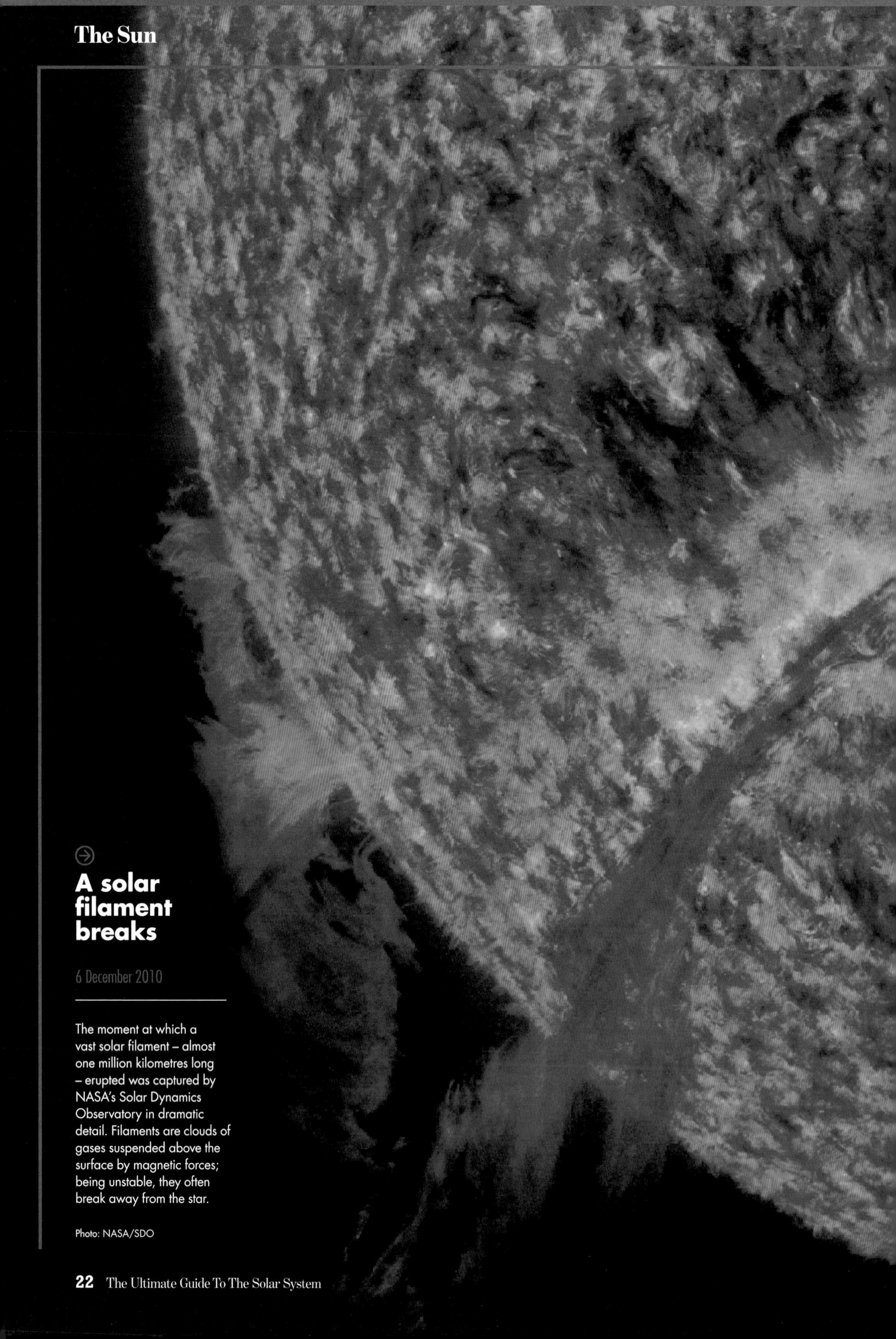

⊕

A solar filament breaks

6 December 2010

_____

The moment at which a
vast solar filament – almost
one million kilometres long
– erupted was captured by
NASA's Solar Dynamics
Observatory in dramatic
detail. Filaments are clouds of
gases suspended above the
surface by magnetic forces;
being unstable, they often
break away from the star.

Photo: NASA/SDO

Coronal mass ejections hitting Earth's atmosphere can cause intense aurorae (northern or southern lights)

Coronal mass ejection

24 January 2007

A powerful burst of solar wind and associated magnetic radiation – a coronal mass ejection – unfurls from the Sun, creating a bulbous, filamentous formation. During periods of high solar activity, CMEs may occur several times a day and, when directed towards Earth, can damage satellites and disrupt radio transmissions. This image was captured by the Solar and Heliospheric Observatory, a craft launched in 1995 to study the Sun in a collaborative project involving NASA and the European Space Agency.

Photo: Courtesy of SOHO/ESA and NASA

Active regions

3 February 2014

The interactions between two large active regions – areas of intense magnetism that appear as sunspots on normal-light images – were captured by the Atmospheric Imaging Assembly on NASA's Solar Dynamics Observatory. Recording two wavelengths of ultraviolet light and enhancing the colour table revealed the high-energy particles spinning along magnetic field lines that arched between the active regions.

Photo: NASA/SDO

Solar flare

13 March 2012

The flare – an intense burst of solar radiation created by the release of magnetic energy – at the far right of this image was classified as medium (M-class), where X is the most powerful and C is the smallest. Captured by NASA's Solar Dynamics Observatory, the image uses 131Å wavelength ultraviolet light to highlight solar activity.

Photo: NASA/SDO

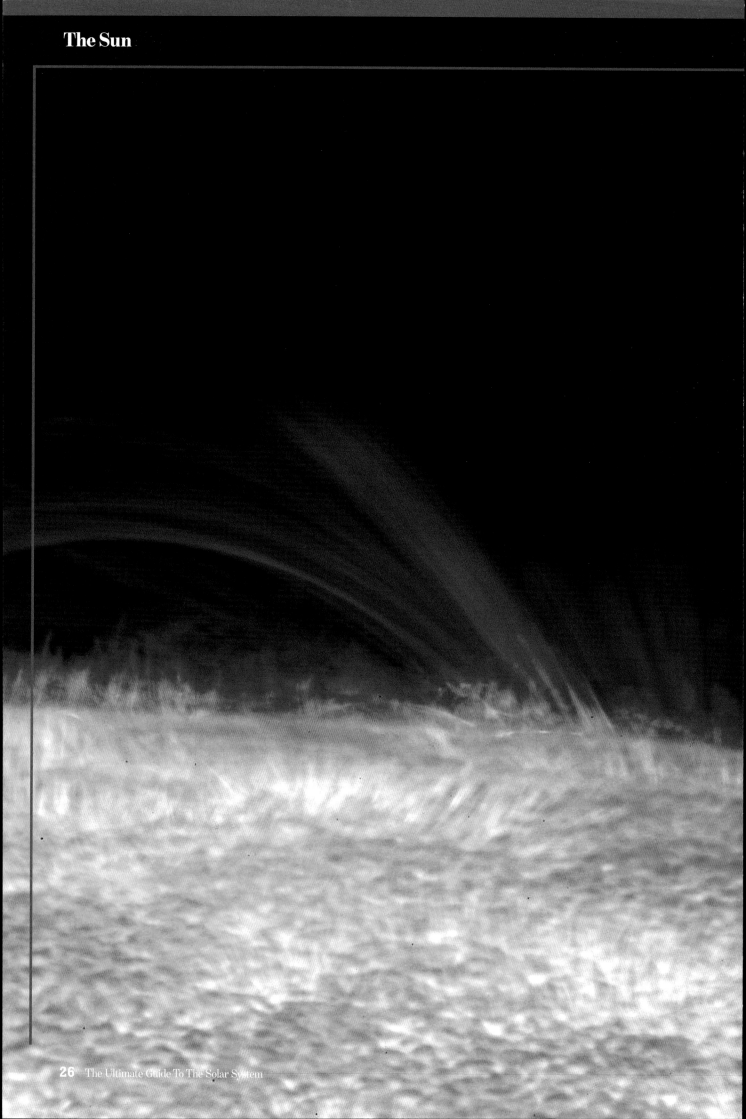

Solar
granulation

11 November 2006

The Sun's photosphere (outer shell) has a grainy appearance. Each of the granules is a convection cell formed by currents of plasma rising in the middle and descending at the edges – but to give an idea of scale, each granule can be 1,000km or more across. This photo, captured by Hinode's Solar Optical Telescope, reveals how hot ionised gases interacting with the Sun's magnetic field stretch outward into the chromosphere from the granules below.

Photo: Hinode JAXA/NASA

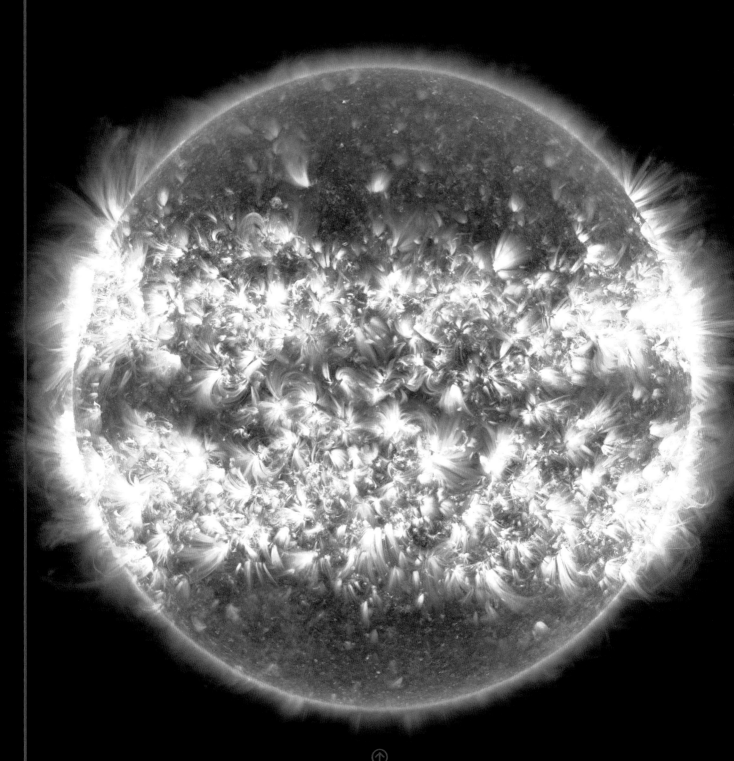

A year of solar activity

16 April 2012–15 April 2013

This composite was created from 25 separate images captured over the course of a year at a wavelength of 171Å – extreme ultraviolet – showing regions of solar activity and the Sun's rotation. The Atmospheric Imaging Assembly of NASA's Solar Dynamics Observatory captures an image of the Sun every 12 seconds in 10 different wavelengths.

Photo: NASA's Goddard Space Flight Center/SDO/S Wiessinger

Solar and Heliospheric Observatory

The Solar and Heliospheric Observatory (SOHO), a collaboration between the European Space Agency and NASA, is a project to study the Sun from its deep core to the outer corona and the solar wind. This photo shows the spacecraft during its construction at Matra Marconi Space facilities. SOHO was launched on 2 December 1995.

Photo: SOHO (ESA & NASA)

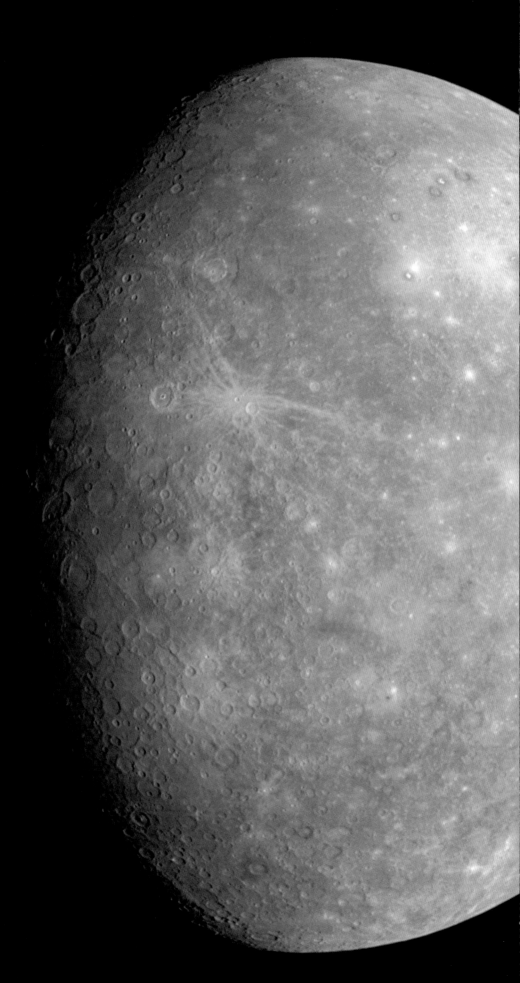

The pitted planet

14 January 2008

To the human eye, the rocky surface of Mercury appears a fairly uniform grey. But by combining images taken with filters affecting different wavelengths, this image – captured by the Wide Angle Camera on NASA's Messenger spacecraft – accentuates the subtle variations in brightness, contrast and hue, and the many craters pocking the surface. The bright patches with rays radiating from them are crushed rock from meteoroid collisions.

Photo: NASA/Johns Hopkins University Applied Physics Laboratory/Carnegie Institution of Washington

Mercury

Mercury is so close to the Sun that its year is a mere 88 days long. The orbit of the smallest planet – only a little larger than our Moon – is highly eccentric, meaning that its distance from our star varies between 45.8 to 70 million kilometres over its year.

The innermost planet is a world of contradictions.

Its proximity to the Sun has resulted in the planet's rotation being near-locked: it spins so slowly that its days are twice as long as its years – so from one sunrise to the next, Mercury will have orbited the Sun twice.

The long days and nights cause the climate of the planet to swing between extremes. During Mercury's day the Sun's heat roasts the surface seven times more fiercely than it does on Earth, raising the temperature to over 420°C. That intense heat and radiation long ago blasted away any vestiges of atmosphere: with no gases to help trap heat near the planet's surface, when night falls the temperature drops to as low as –180°C. Ⓢ

Crater in the Caloris Basin

28 October 2011

This finely etched (and as-yet unnamed) crater lies within the Caloris Basin, an impact crater itself over 1,500km across, visible as a dark patch at top right on the picture to the left. This image was also taken by NASA's Messenger.

Photo: NASA/Johns Hopkins University/ Carnegie Institution

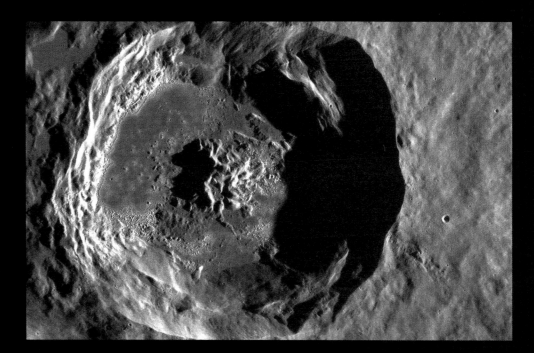

→

Polar darkness

Mercury's north pole, unlike Earth's, is in persistent shadow, and extremely cold. The red patches in this image depict regions that have never been photographed in light by Messenger; yellow areas – seen more clearly on the image at bottom right – show where deposits believed to be water ice frozen in the darkness have been identified by Earth-based radar.

Photo: NASA/Johns Hopkins University Applied Physics Laboratory/Carnegie Institution of Washington/National Astronomy and Ionosphere Center, Arecibo Observatory

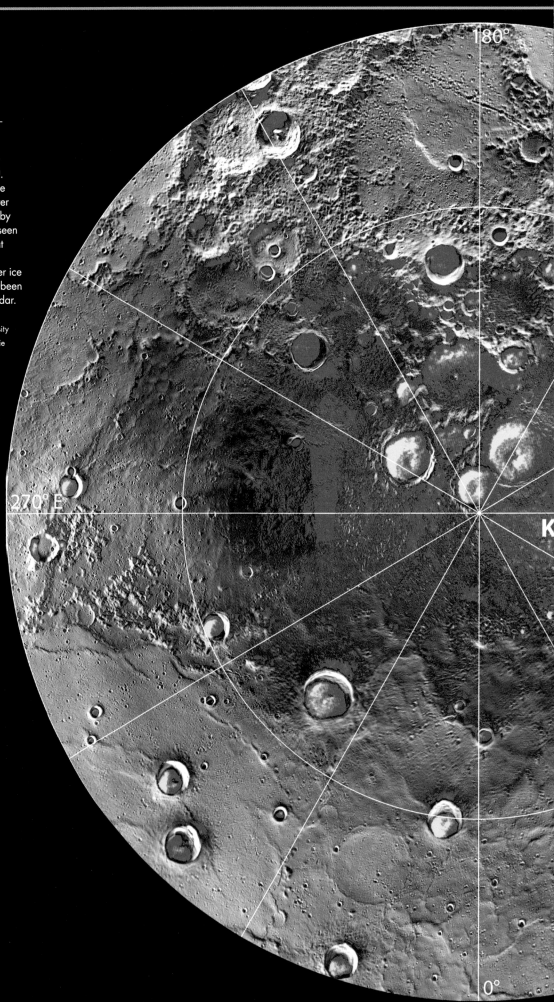

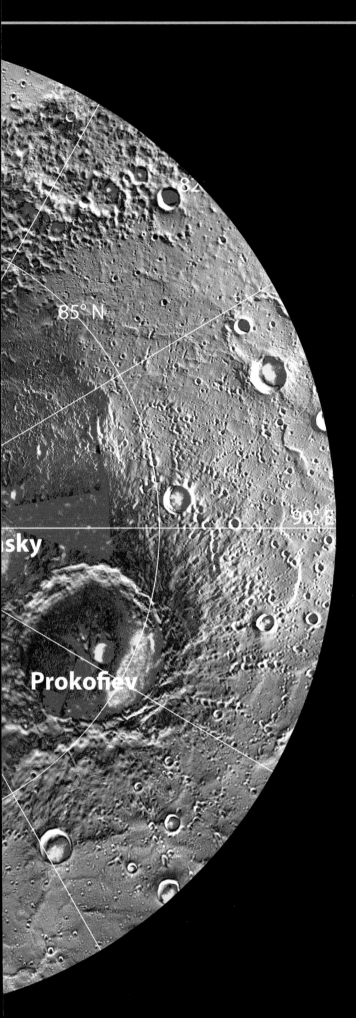

85° N

Prokofiev

Orbital mosaic

8 April 2013

This image of the northern polar region was created from thousands of photos captured by Messenger's Mercury Dual Imaging System during more than two years of operation.

Photo: NASA/Johns Hopkins University Applied Physics Laboratory/Carnegie Institution of Washington

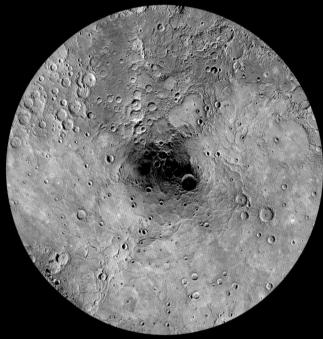

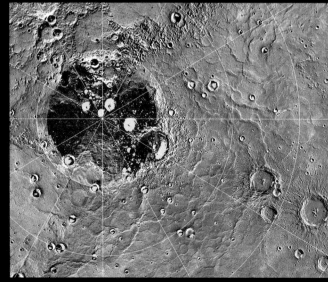

Volcanic vents

23 August 2013

Not all of the topographic features on Mercury's surface were created by meteoroid collisions This beautiful image, captured by Messenger's Wide Angle Camera, shows a bright area comprising what is believed to be pyroclastic material surrounding the vent that produced it, possibly a few billion years ago. The irregular-shaped formation at bottom right is an older vent.

Photo: NASA/Johns Hopkins University Applied Physics Laboratory/Carnegie Institution of Washington

Pyroclastic vents

2 August 2013

The yellowish patches on this image depict pyroclastic vents. Though commonly seen across the planet, this is one of the largest such clusters, stretching across nearly 10° of Mercury's surface. These vents are believed to have been the source of explosive eruptions driven by volcanic gases.

Photo: NASA/Johns Hopkins University Applied Physics Laboratory/Carnegie Institution of Washington

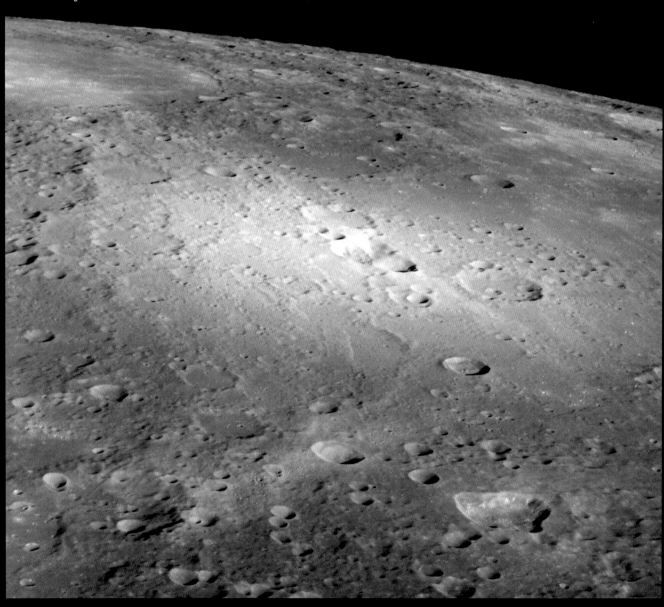

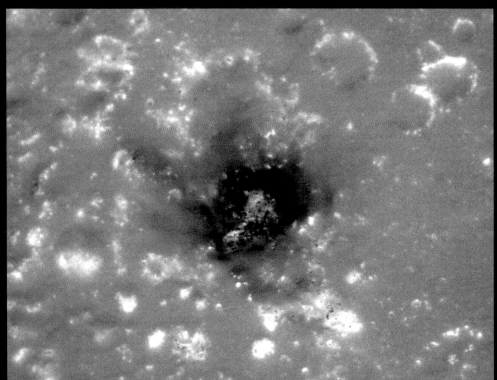

⊖ Dark matter

5 May 2013

The irregular-shaped depression at the centre of this image is believed to be a pyroclastic vent – though an unusual one, surrounded by a dark area of low-reflectance material (LRM) of unknown composition, rarely seen at other vents. This one is sited at the southern rim of the Caloris Basin.

Photo: NASA/Johns Hopkins University Applied Physics

⊕

The colour difference

22 February 2013

The impact that the use of colour filters makes on image quality is clearly visible in these two composite images, both comprising thousands of individual shots. The version at left combines monochrome images, while that at right is made up of red-, green- and blue-filtered images. Variations in surface topography and composition are far more clearly depicted – though we still know little of the minerals that make up the planet's surface.

Photos: NASA/Johns Hopkins University Applied Physics Laboratory/Carnegie Institution of Washington

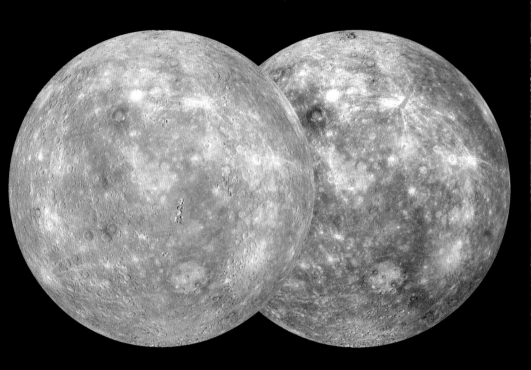

⊖

Crater Bartók

26 April 2013

Bartók, at the top left of this image, is a 116km-wide complex crater enclosing a central peak. It is surrounded by several smaller craters formed by ejecta – material expelled during the impact that formed the main crater. The lighter patch towards the right of the image is a younger crater formed by a relatively recent impact, indicated by the more visible radiating streaks of ejecta.

Photo: NASA/Johns Hopkins University Applied Physics

⊕

Hues of Mercury

Global image data from Messenger was compiled to create this vivid colour image representing Mercury's physical, mineralogical and chemical regions. The large sandy-coloured patch at top right is the Caloris Basin, flooded with volcanic lava after the impact that created the initial crater.

Photo: NASA/JHU Applied Physics Lab/ Carnegie Inst. Washington

Discovery Rupes

6 October 2008

The ragged line stretching diagonally in from the top left of this image is called Discovery Rupes, a huge escarpment 650km long and 2km high. It's thought that such thrust faults – there are many, of different lengths, on Mercury – were formed as the planet's core cooled and contracted, fracturing the surface. This photo, captured by Messenger's Narrow Angle Camera, shows that the scarp has deformed a large impact crater, itself over 100km across.

Photo: NASA/Johns Hopkins University Applied Physics Laboratory/Carnegie Institution of Washington

Messenger begins its mission to Mercury

3 August 2004

Smoke billows as the Boeing Delta II rocket, with the
Messenger spacecraft on top, blasts off from Cape Canaveral.
In January 2008, Messenger became the second spacecraft
(after Mariner 10 in 1974) to reach Mercury. By March 2013,
it had achieved 100 per cent mapping of the planet.

Photo: NASA

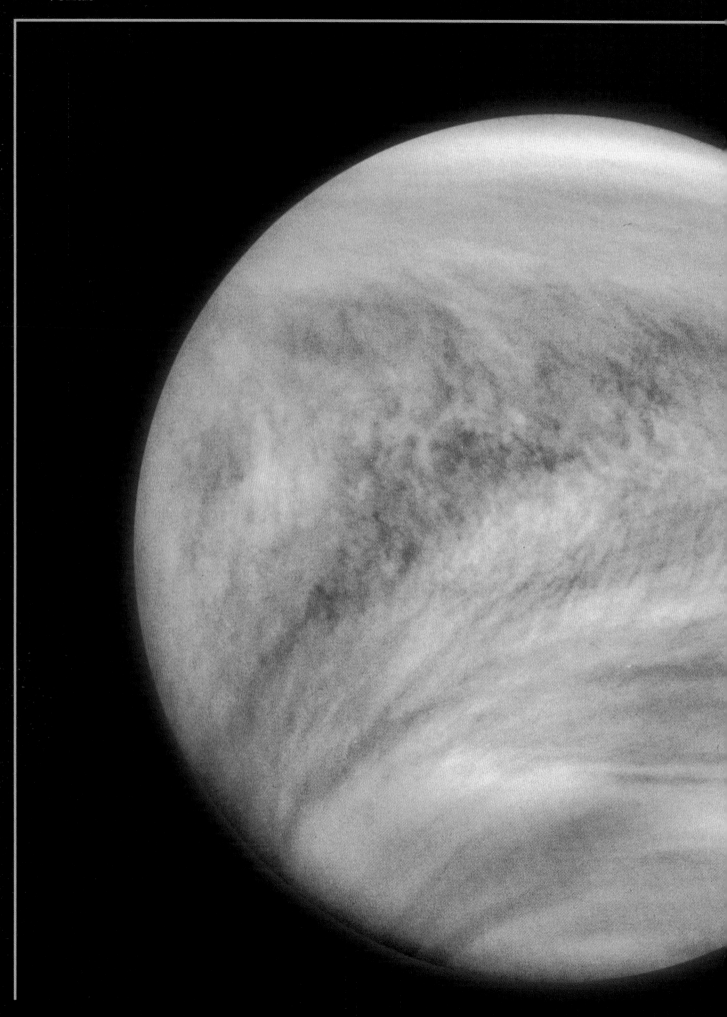

Venus

Despite being named for the Roman goddess of love, the planet Venus is anything but romantic. Just 45 million kilometres from our planet's orbit, and with a mass 80 per cent that of the Earth's, Venus is our nearest neighbour in terms of both distance and size – yet our two worlds are vastly different.

The atmosphere of Venus is the densest of any of the terrestrial planets – air pressure at its surface is 92 times that on Earth. It is also highly toxic, composed of 96 per cent carbon dioxide. Such a high concentration of greenhouse gases caused a dramatic build-up of heat: temperatures at the surface can reach 462°C.

Thick clouds mask the terrain below but we do have a few images of the surface of Venus – the first taken on another planet – from Soviet Venera probes. It seems Venus was once not so different from Earth, and may even have had oceans before its deadly atmosphere created the desolate world we see today. **S**

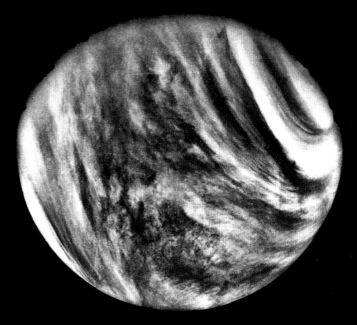

← ↑

Cloud formations

26 February 1979 (left)
5 February 1974 (above)

Clouds of sulphuric acid cloak Venus. These pictures – an ultraviolet image taken by the Pioneer Venus Orbiter, left, and a colour-enhanced shot aiming to replicate what the human eye would see, taken by the Mariner 10 probe – show formations created by winds reaching 360km/h at high altitudes.

Photos: NASA

Radar topography

The Venera 15 and 16 probes launched by the Soviet Union in 1983 mapped Venus using surface imaging radar equipment, revealing features such as ridges and canyons, as seen in this image. It's believed that these formations are the result of tectonic activity, though we still know relatively little about the planet's surface and the forces that moulded it.

Photo: Detlev Van Ravenswaay/ Science Photo Library

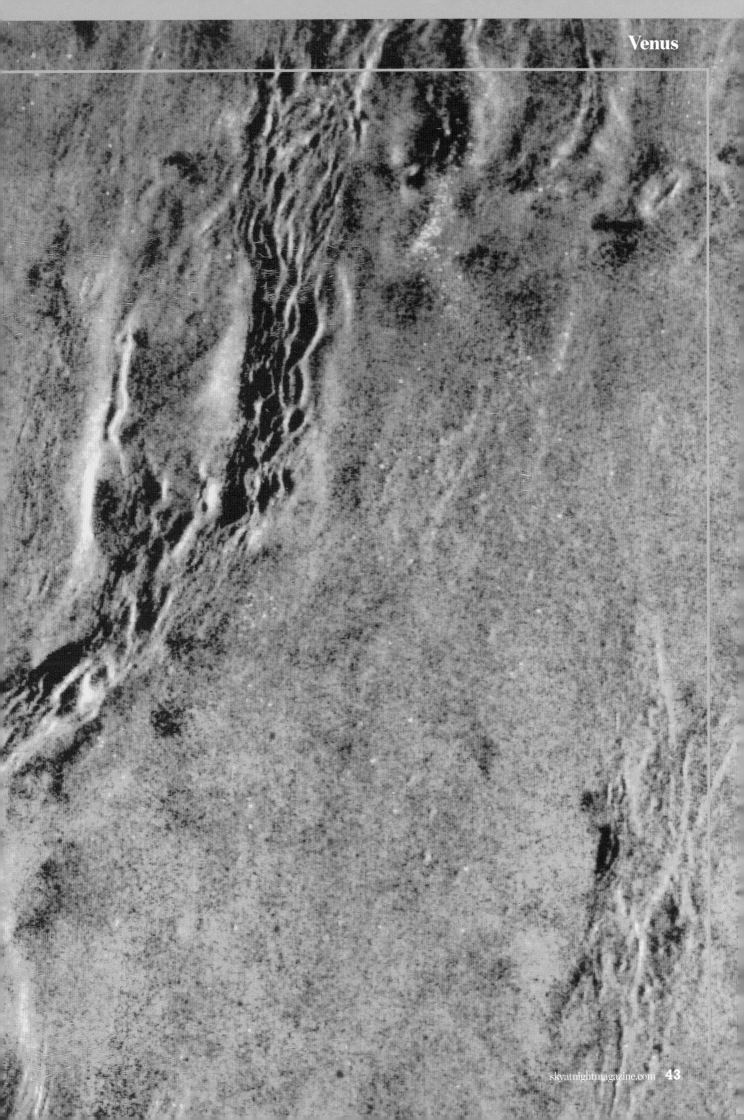

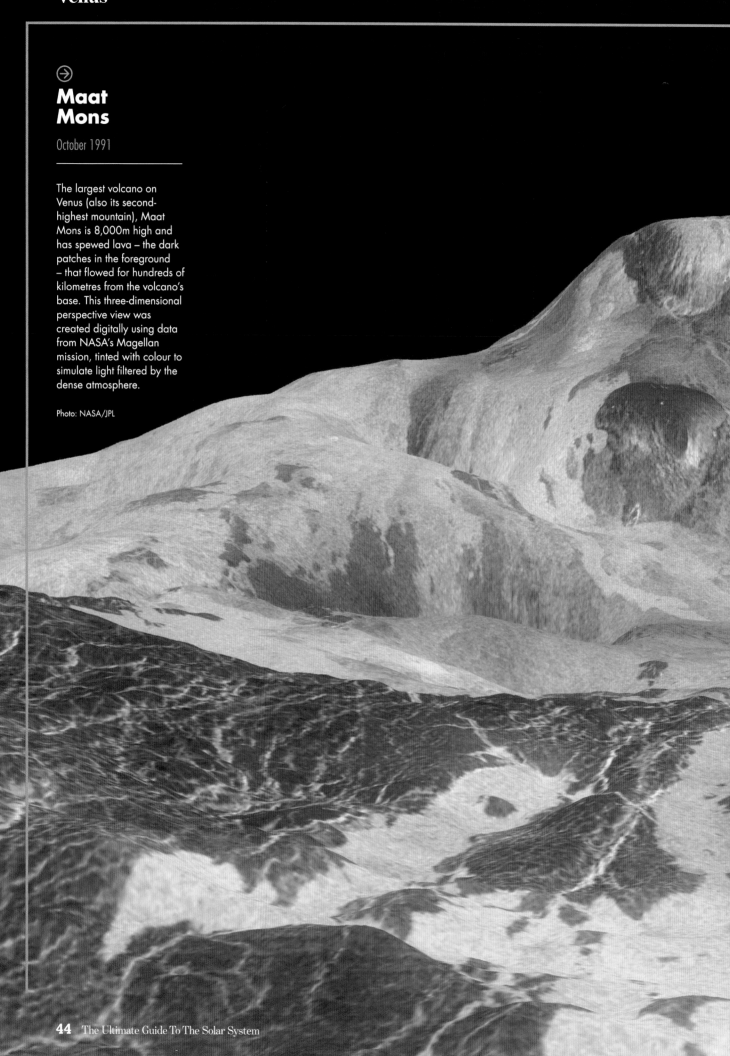

Maat Mons

October 1991

The largest volcano on Venus (also its second-highest mountain), Maat Mons is 8,000m high and has spewed lava – the dark patches in the foreground – that flowed for hundreds of kilometres from the volcano's base. This three-dimensional perspective view was created digitally using data from NASA's Magellan mission, tinted with colour to simulate light filtered by the dense atmosphere.

Photo: NASA/JPL

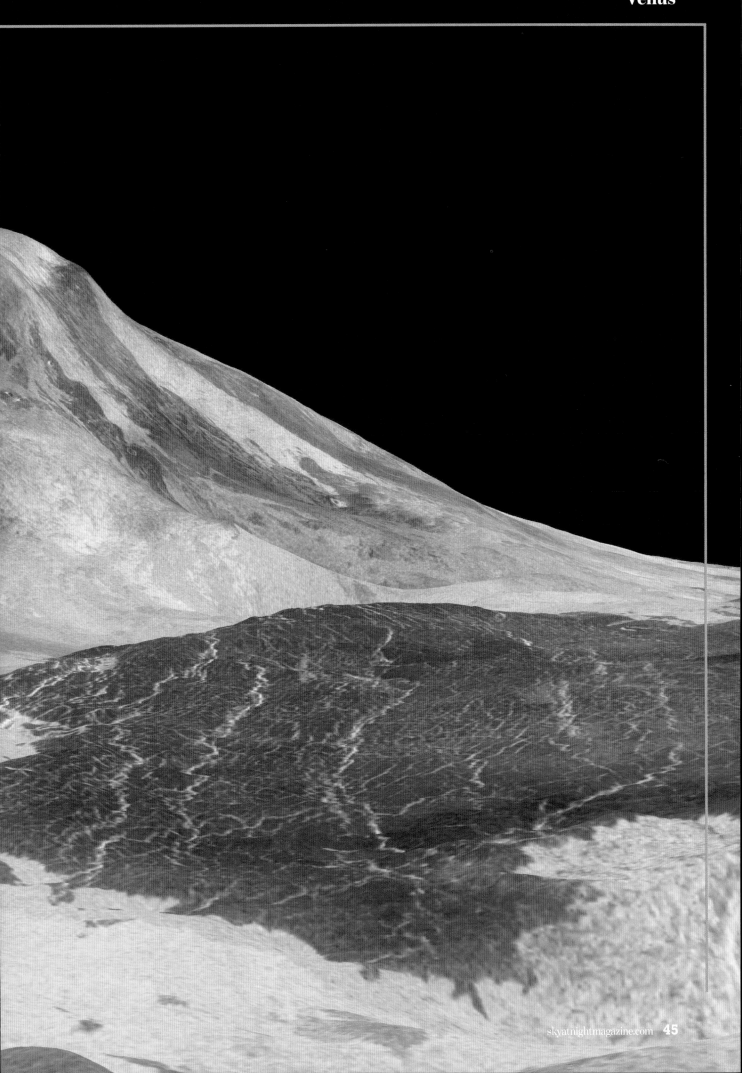

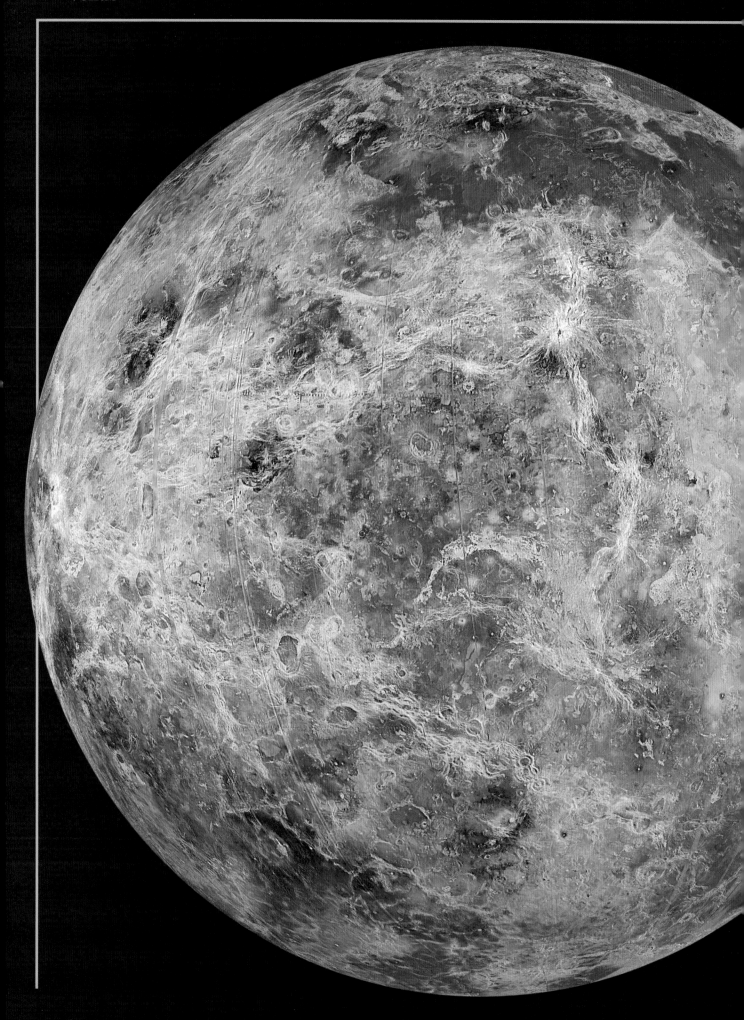

Terrain map

16 May 2012

Though recently created, this map of the surface of Venus incorporates data accumulated from over a decade of radar imaging – much of it by Magellan, but some from Earth-based Arecibo Observatory and other space missions. Colours indicate differences in elevation, depicting steep-sided domes and high-altitude regions, and the map might suggest the presence of mineral-rich felsic rocks.

Photo: NASA/JPL/USGS

Gula Mons

5 March 1991

A computer-generated image of Gula Mons, a 3km-high volcano, depicts the double cone. This picture was again created using data from Magellan, and tinted based on colours detected by the Soviet Venera 13 and 14 spacecraft.

Photo: NASA/JPL

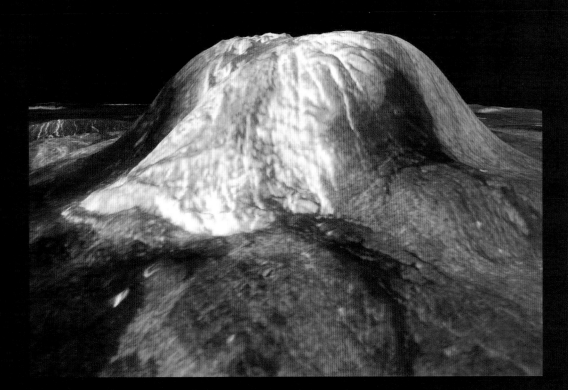

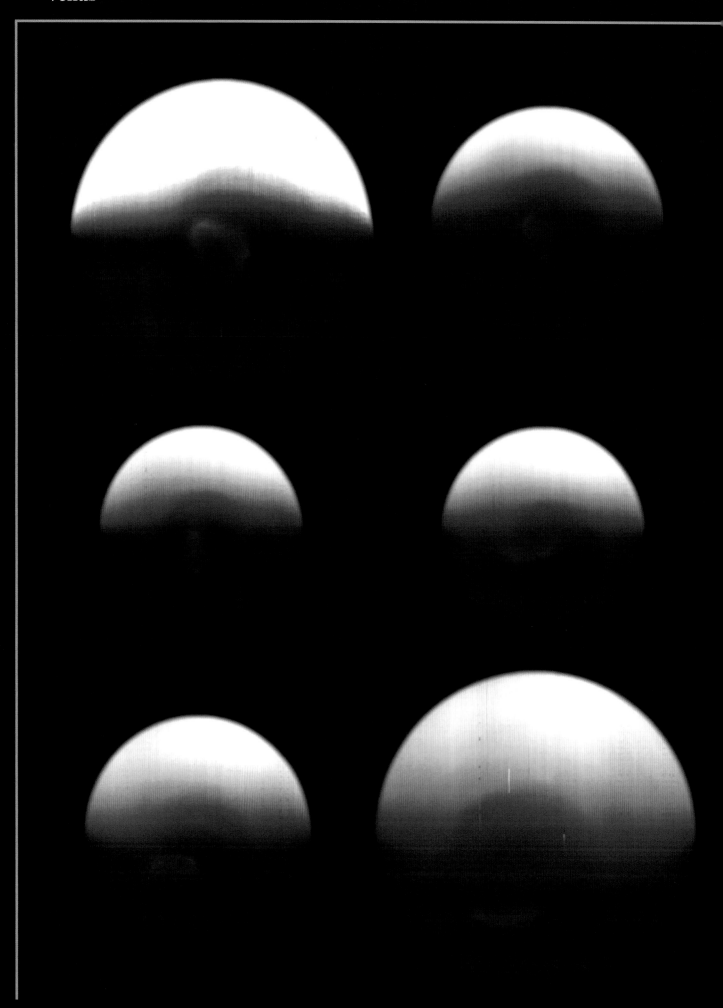

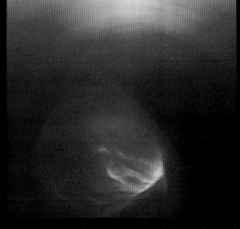
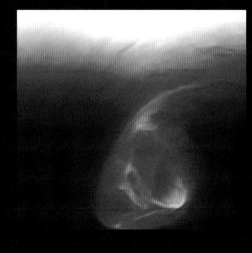
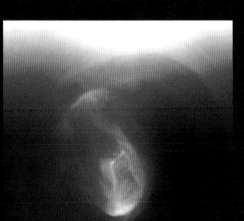

Southern polar vortices

12–19 April 2006 (far left)
August 2007 (left)

Permanent vortices in the clouds over both of Venus's poles change rapidly over a matter of days or even hours. These infrared images, captured by the Visible and Infrared Thermal Imaging Spectrometer during multiple orbits of the European Space Agency's Venus Express craft, show variations in the shape of the southern polar cyclone, which is fed by the rapid movement of the planet's atmosphere – rotating 60 times faster than Venus itself.

Photos: ESA/VIRTIS-VenusX/INAF-IASF/
Obs. de Paris-LESIA (A Cardesin, Moinelo)

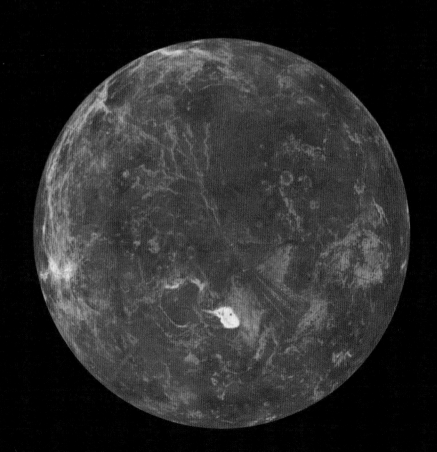

Global view of the northern hemisphere

26 May 1993

This image, produced using radar mapping from the Magellan mission plus data from the Venera 13 and 14 landing craft, Pioneer-Venus Orbiter and Arecibo Observatory, depicts the planet's main surface features. The bright yellow shape in the lower central part of the image is Maxwell Montes, the highest mountain massif on Venus, which rises to an altitude of 11km above the planet's average elevation.

Photo: NASA/JPL

Sapas
Mons

28 May 1991

The massive shield volcano
known as Sapas Mons was
mapped by Magellan, clearly
showing radial lava flows that
appear to have emanated
from flank eruptions rather
than the twin flat-topped
mesas at the volcano's 1.5km-
high summit. The volcano,
here mapped using imaging
radar data from Magellan,
is about 400km across.

Photo: NASA/Science Photo Library

Photo: NASA

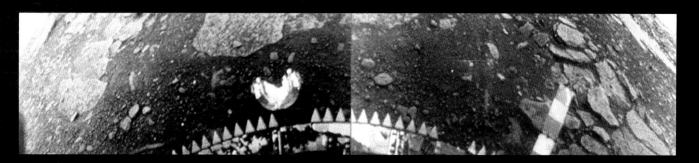

⊕

Soviet Venera 13 lander

October 1981

The Venera 13 mission was launched on 30 October 1981; the descent
craft (pictured top) separated from the main vessel on
1 March 1982 and plunged into the atmosphere of Venus. It survived
the landing and harsh environment – a surface temperature of 457°C
and pressure of 84 Earth atmospheres – for 127 minutes, long enough
to capture and transmit an image panorama (above)and fluorescence
spectrometric analysis of a soil sample.

Photo: NASA

Mars

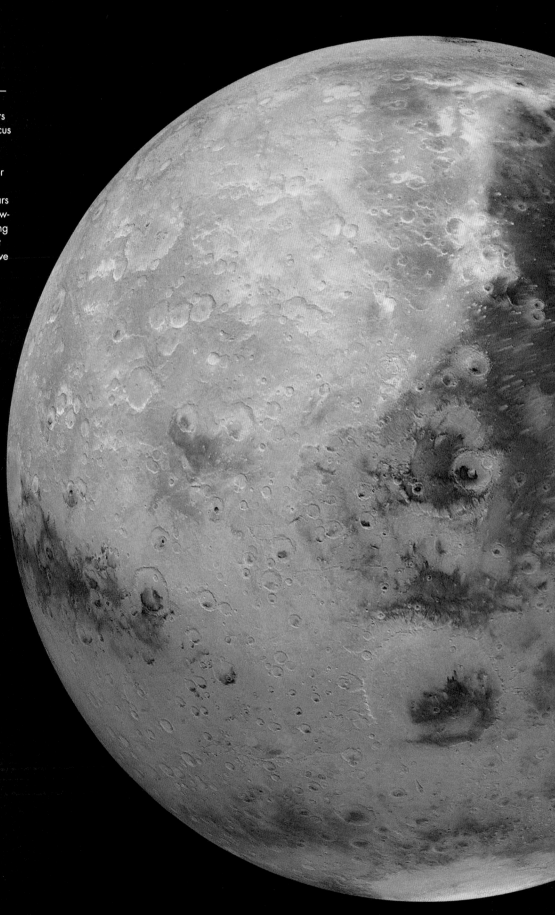

⊕→

The majestic Red Planet

1980

Instantly recognisable, Mars continues to be a major focus of space exploration. The dark area on the right side of this image is Syrtis Major Planum, a 1,000km-wide feature revealed by the Mars Global Surveyor to be a low-lying shield volcano. Lacking the fine iron-oxide dust that gives the planet its distinctive hue, basaltic volcanic rock is exposed instead.

Photo: NASA/MDIM/Jody Swann/ Tammy Becker/Alfred McEwen/ US Geological Survey, Arizona

Mars

With a diameter half that of Earth, Mars is the Solar System's second smallest planet, but is probably the one that has most captured the human imagination. Until Mariner 4 sent back the first close-up images of Mars in the 1960s, many believed it might harbour life.

A few billion years ago, Mars probably didn't look that different from our own planet. Only another 75 million kilometres farther away from the Sun than the Earth, it was warm enough to maintain liquid oceans on most of the surface and its short day of 24 hours and 40 minutes helped to keep the temperature constant across the planet's surface.

Now things are very different. At some point the planet was stripped of its atmosphere and, with it, most of the water. Though some remains frozen into the soil, this is now a barren, dead landscape.

However, many believe that, in its more pleasant days, Mars might once have been home to more than rocks and dust. Dozens of missions have been sent to the planet, hoping to find signs of life. $\circledS$

Olympus Mons volcano

22 June 1978

Looking like a blister on the surface of Mars, this is another shield volcano – Olympus Mons, found in the western hemisphere. It's also the second tallest mountain in the Solar System; standing more than 21,000m, it's more than double the height of Mount Everest.

Photo: NASA/MDIM/Jody Swann/Tammy Becker/Alfred McEwen/US Geological Survey, Arizona

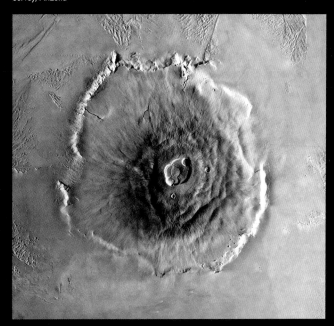

Pavonis Mons

2006

These striking geological features, found on the flank of the shield volcano Pavonis Mons, are believed to be channels formed by hot lava that, when it cools, creates a hard crust. Hot lava continues to run through these tubes before the channels collapse, leaving intriguing depressions on the planet's surface.

Photo: ESA/DLR/FU Berlin
(G Neukum)

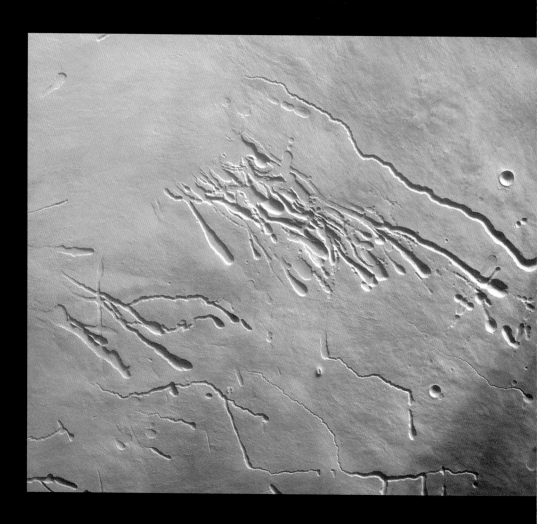

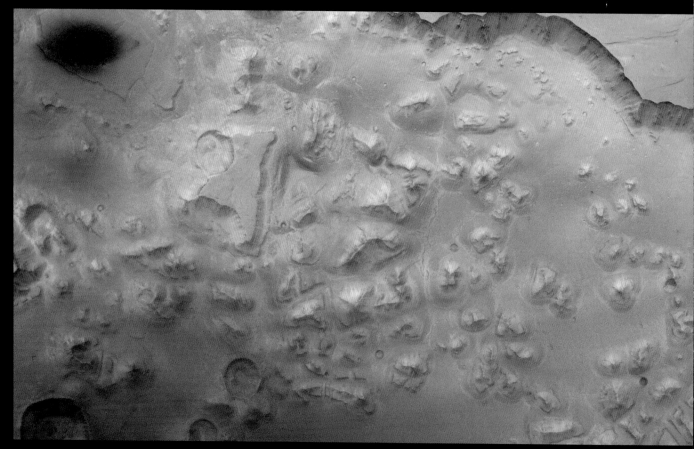

Valles Marineris canyon

The largest canyon in the Solar System, Valles Marineris scars the face of Mars. Eight kilometres deep at its most incursive point, the valley boasts more than four times the depth of Arizona's Grand Canyon. The image to the left was put together using more than 100 pictures taken by the two Viking Orbiters in the 1970s.

Photos: Left – Viking Project/USGS/NASA
Below – ESA/DLR/FU Berlin (G Neukum)

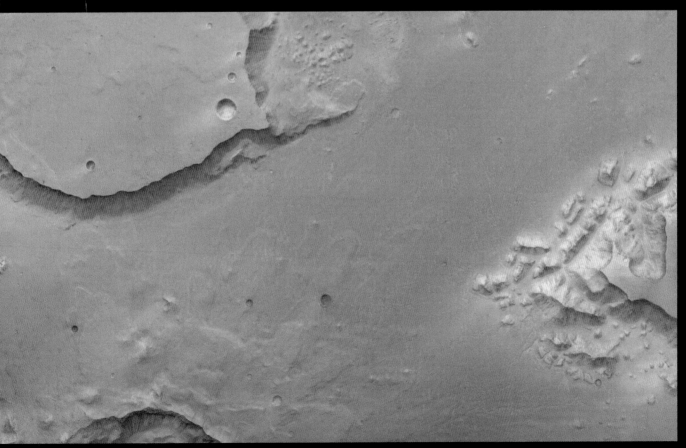

⊕

Victoria crater

3 October 2006

Located at Meridiani Planum near Mars's equator, the 800m-wide Victoria crater looks beautiful in this image taken by NASA's Mars Reconnaissance Orbiter in 2005. The pattern inside the crater is an extended area of sand dunes.

Photo: NASA/JPL/University of Arizona

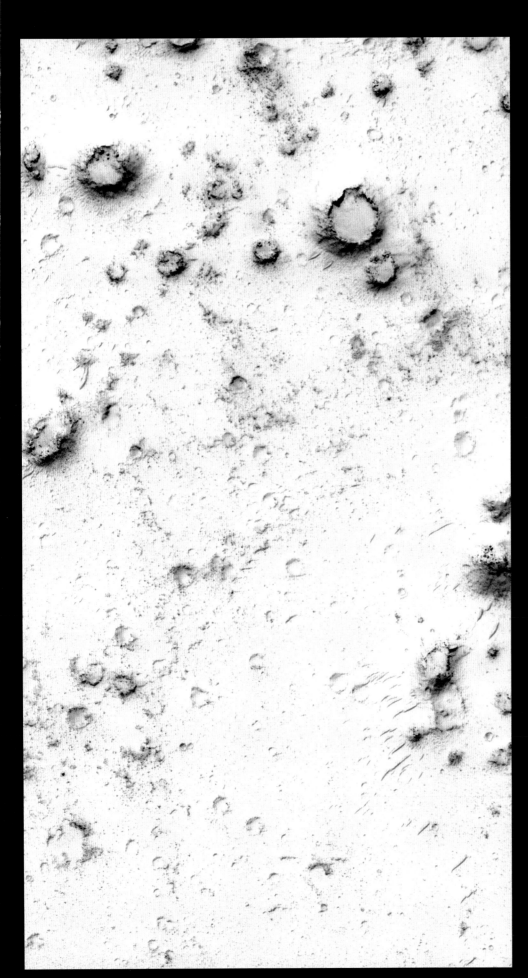

Frosty craters

Most of the craters on Mars are the result of asteroids or comets colliding with the Red Planet. These craters, though, found in the Tartarus Montes mountain range, are believed to have been caused by steam explosions that may have occurred when hot lava has flowed over icy terrain.

Photo: NASA/JPL-Caltech/University of Arizona

➔

Olympus Mons in cloud

This image, captured by one of the Viking orbiters, shows the giant volcano Olympus Mons shrouded in cloud. The clouds – most likely composed of water-ice – sit a full 8km below the volcano's peak.

Photo: NASA/Science Photo Library

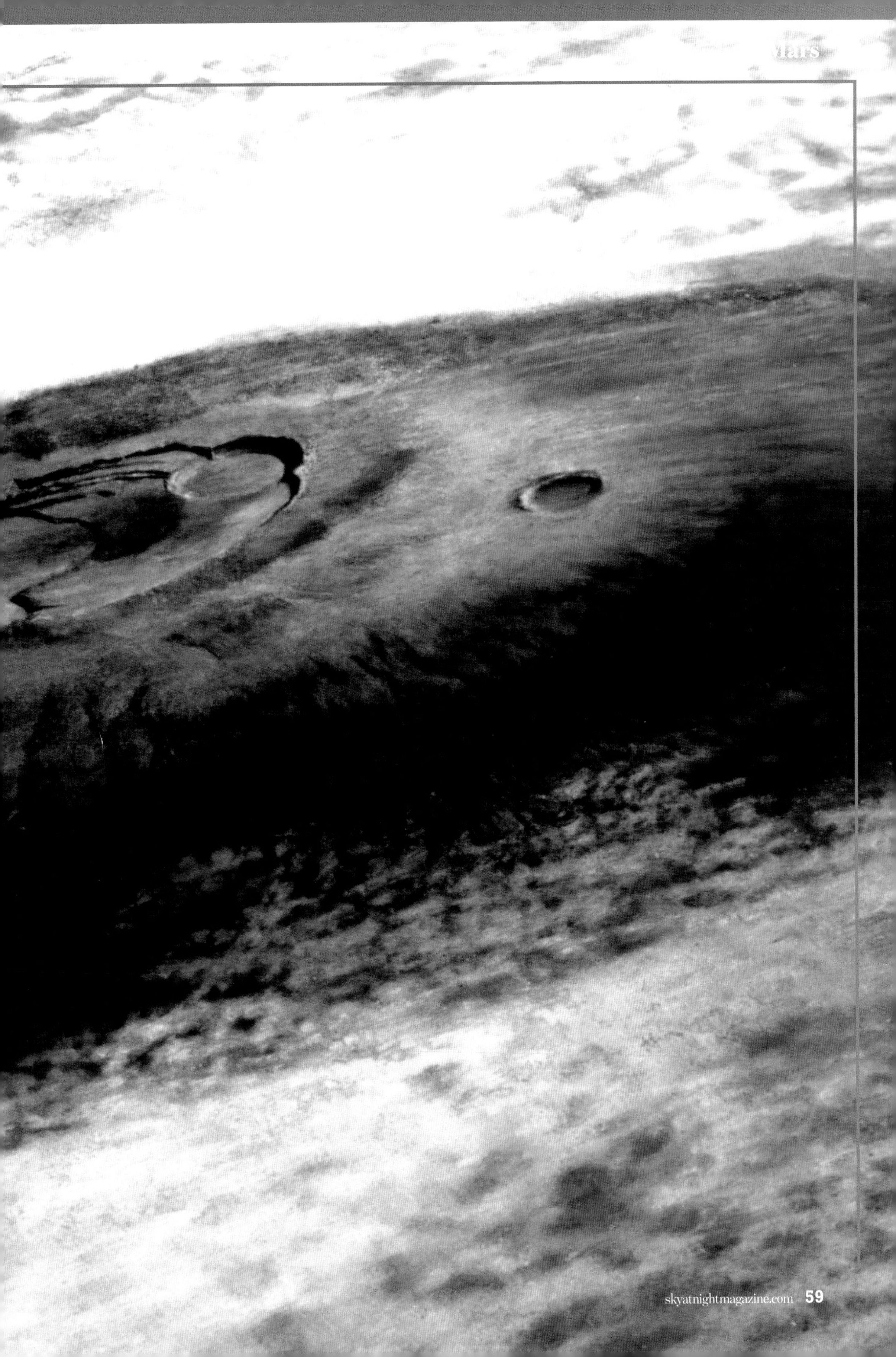

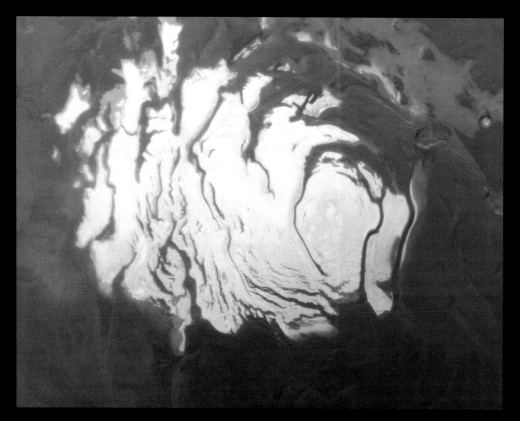

The south polar cap
17 April 2000

Taken by the Mars Global Surveyor's Mass Orbiter Camera, the south polar cap is at its smallest (about 420km wide) as Mars's southern hemisphere hits the height of its summer. By June, the frost cap will be increasing again as the south heads towards winter, when this region will be completely covered in ice.

Photo: NASA/JPL/Malin Space Science Systems

The atmosphere of Mars

The thin atmosphere on Mars is visible on the horizon in this photograph. The planet's atmosphere is 96 per cent carbon dioxide, although trace elements of methane found in recent years have got scientists salivating over the prospect that its presence could confirm that there was life here once.

Photo: NASA

⊕

Spheres of mystery

6 September 2012

A field of tiny spherical objects is found on the western rim of Endeavour crater. These spherules – nicknamed 'blueberries' – measure around 3mm in diameter and were discovered by the Mars exploration rover Opportunity, whose Microscopic Imager took the four photographs that make up this mosaic image. Further study of these spherules may enhance understanding of past environmental conditions on Mars.

Photo: NASA/JPL-Caltech/ Cornell University/USGS/ Modesto Junior College

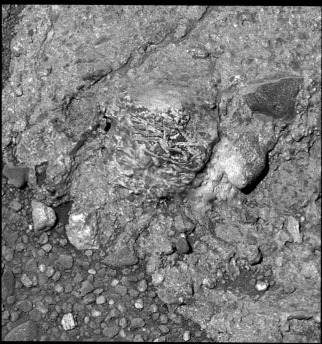

The findings of Curiosity Mars rover

Since it landed on Mars on 6 August 2012, NASA's Curiosity Mars rover has improved understanding of the planet immeasurably, sending back images both colourful and phenomenally detailed. Photographs such as these two of the surface geology continue to demystify this ever-alluring planet.

Photos: NASA/JPL-Caltech/LANL/CNES/ IRAP/LPGNantes/CNRS/IAS/MSSS

Dingo Gap

28 January 2014

In January 2014, NASA's Curiosity Mars rover approached 'Dingo Gap', a shallow valley between two gentle escarpments. The view here combines images taken by the left-eye camera of the Mastcam on board Curiosity.

Photo: NASA/JPL-Caltech/MSSS

Mars

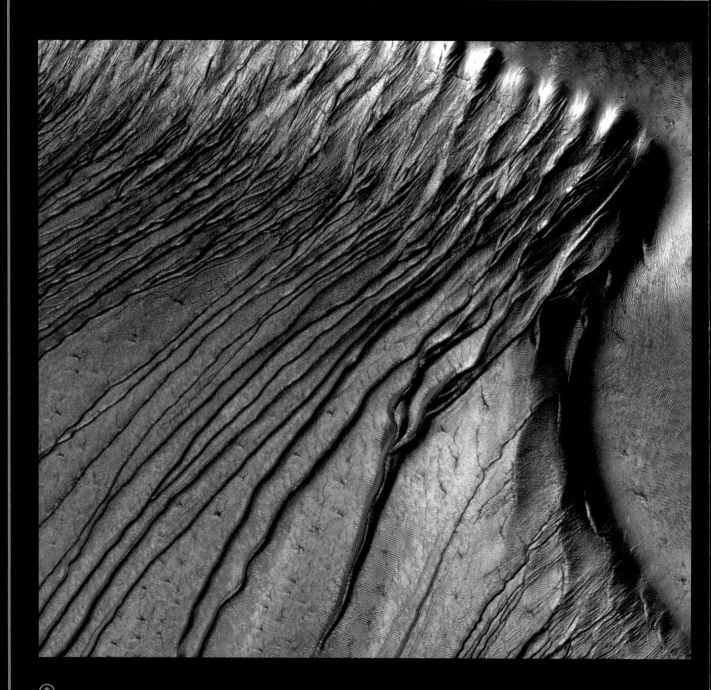

Sand dunes in the Russell crater

This is an extraordinary image of sand dune ridges in the Russell crater taken by Mars Reconnaissance Orbiter in November 2013. The dark areas are where the dust layer has been blown away, exposing the darker basalt layer underneath. The whiter patches represent carbon dioxide frost.

Curiosity Mars rover

3 February 2013

Curiosity sends home a 'selfie' – a composite of dozens of shots taken by the rover's Mars Hand Lens Imager. The spacecraft was launched in November 2011 on a mission to study the planet's geology at close quarters.

Photo: NASA/JPL-Caltech/MSSS

The surface of asteroid Vesta

24 July 2011

This dramatic image of the giant asteroid Vesta was captured by NASA's Dawn spacecraft at a distance of 5,200km. As well as investigating Vesta, Dawn was launched to study the dwarf planet Ceres, where it's scheduled to arrive in February 2015. This image of Vesta shows parallel grooves running around its equator, along with innumerable craters of differing sizes.

Photo: Jet Propulsion Laboratory

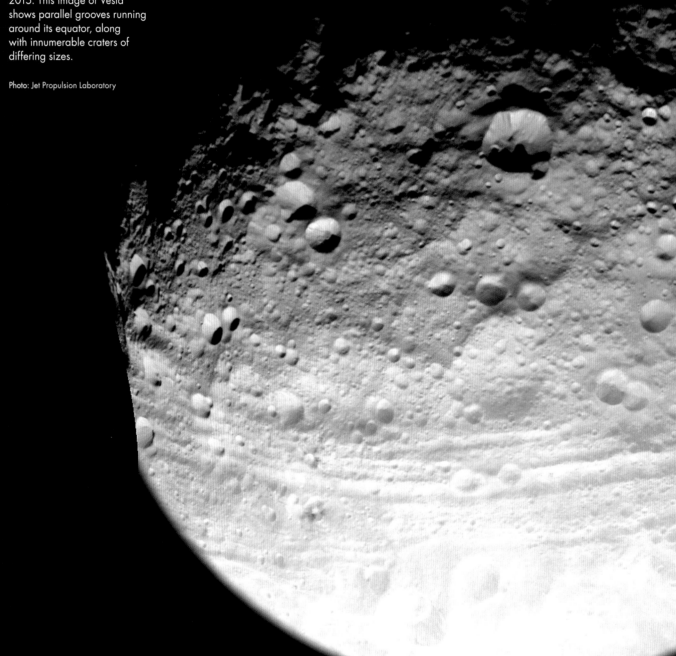

The asteroid belt

Spanning the gap between Mars and Jupiter is a ring of rocky asteroids, 300 million to 600 million kilometres from the Sun. Once destined to be an Earth-sized planet, the potential world was pulled apart by Jupiter, scattering most of its material into deep space, leaving only 0.01 per cent of the mass behind.

Ceres was the first asteroid discovered, in 1801; it makes up a third of the belt's mass. Since then, hundreds of thousands more have been found; it's suspected that there are more than a million asteroids with a diameter greater than 1km.

Despite the large number of asteroids, the region is mostly empty space, with as much as a million kilometres between individual asteroids. They are so far apart that an astronaut standing on one would need a pair of binoculars if he wanted to see to the next one.

A few missions have landed on asteroids, and there are even plans to begin mining them for their metals. With iron-rich asteroids estimated to contain $700 million trillion of mineral wealth, these plans might not be as outlandish as they first seem. Ⓢ

⊕

Ceres the dwarf planet

23 January 2004

The largest body found in the asteroid belt, Ceres accounts for a third of the belt's mass. Its surface area is around 12 times that of the UK.

Photo: NASA/ESA/J Parker (Southwest Research Institute), P Thomas(Cornell University), L McFadden(University of Maryland,College Park)

Inside Eros's craters

14 June 2000

Eros, a near-Earth asteroid 33km long and 13km wide, was studied extensively by NASA's NEAR Shoemaker probe, which also took multiple images of the asteroid's large crater. This false-colour image shows the crater's regolith – the loose material covering Eros's rock. The redder shades show regolith that's been chemically altered after contact with solar winds and modest impacts with other bodies.

Photo: NASA/JPL/JHUAPL

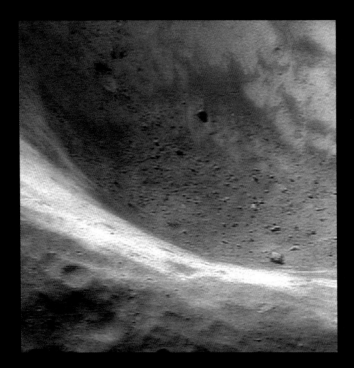

The behaviour of an active asteroid

10 & 23 September 2013

Six comet-like tails trail from this body in the asteroid belt, designated P/2013 P5. Both images were taken by the NASA Hubble Space Telescope. The first, on the left, was taken on 10 September 2013. To the surprise of those monitoring it, the second, captured 13 days later, shows the asteroid to have spun around, possibly due to radiation pressure.

Photos: NASA/ESA/D Jewitt (UCLA)

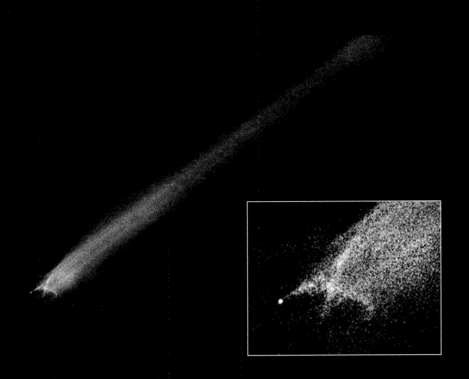

Head-on asteroid collision

29 January 2010

It might look like a comet, but it isn't. NASA's Hubble Space Telescope suggested this object, called P/2010 A2, was actually the product of a head-on collision between two asteroids travelling at around 5km per second approximately 140 million kilometres from Earth. A blue colour filter has been used to enhance the object's finer details.

Photo: NASA

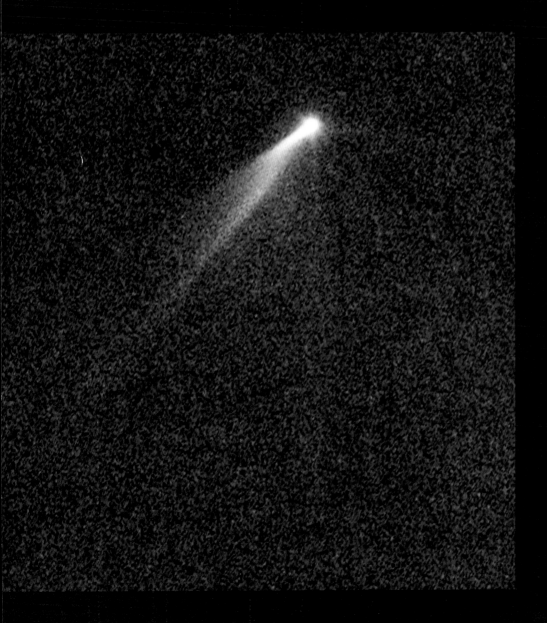

Vesta's Antonia crater

September–October 2011

This stunning, colourised composite image shows Antonia, a 17km-wide crater in the southern hemisphere of the giant asteroid Vesta. The light blue represents a grain-like material excavated from the asteroid's lower crust, while the dark blue shows the shadows cast by the crater's southern rim. The images combined here were taken by the Dawn probe.

Photo: NASA/JPL-Caltech/UCLAMPS/ DLR/IDA

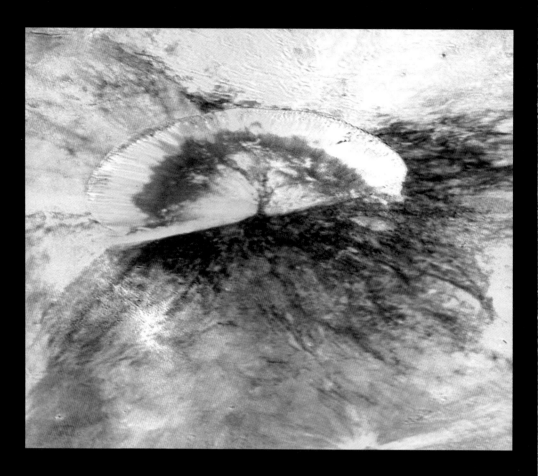

The Dawn probe orbits asteroid Vesta

An artist's impression of the NASA probe Dawn orbiting Vesta. Having been launched in September 2007, the spacecraft made its initial orbit of the asteroid in July 2011, leaving its gravitational pull 14 months later, bound for Ceres.

Photo: NASA/JPL-Caltech

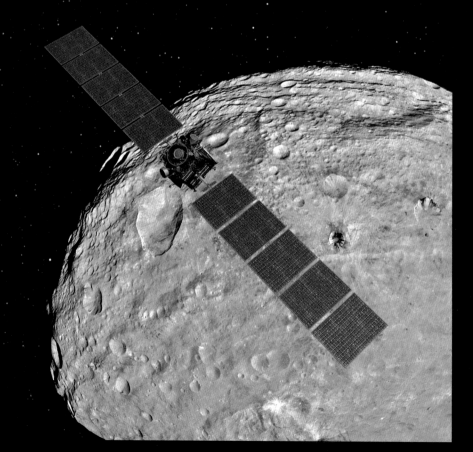

Pioneer 10 heads for the asteroid belt

3 March 1972

The Pioneer 10 space probe is launched aboard an Atlas-Centaur launch vehicle. As it travelled towards Jupiter, on 15 February 1973 it became the first spacecraft to traverse the asteroid belt. The last signal received from Pioneer 10 was on 23 January 2003, by which time it had travelled 12 billion kilometres from Earth.

Photo: NASA

Jupiter from a distance

December 2000

The colours of Jupiter are captured by the Cassini spacecraft in this stunning image showing spectacular detail and structure in the gas giant's stormy atmosphere. This picture was taken in December 2000 as the spacecraft passed by Jupiter on its way to rendezvous with Saturn.

Photo: Thinkstock

Jupiter

Jupiter is the first of the gas giants – and the most spectacular. At 140,000km across, it is large enough to house the rest of the Solar System's planets twice over.

With a mass 318 times that of the Earth, Jupiter's gravitational pull affects all of the planets and bodies in the Solar System. During the system's formation, Jupiter played a pivotal role by sucking up material that could have gone on to form other planets.

Despite its size, Jupiter is the planet with the shortest day, taking less than 10 hours to complete one rotation. However, measuring the day's length has been a challenge; being a gas giant, it doesn't have a solid surface with static features we can track. Instead, Jupiter is composed of many layers of gases that move independently. The only layer we can see is the upper 50 km, a huge swirl of currents that is home to Jupiter's best-known feature, the Great Red Spot, a storm the size of the Earth that's been raging for centuries. ⓢ

The Brown Barge

2 March 1979

Also captured by Voyager 1, this image shows a long, brown oval on Jupiter's surface known as the Brown Barge. Measuring up to 10,000km across, it's believed that the oval represents an opening in the upper cloud deck that reveals the planet's deeper layers.

Photo: NASA

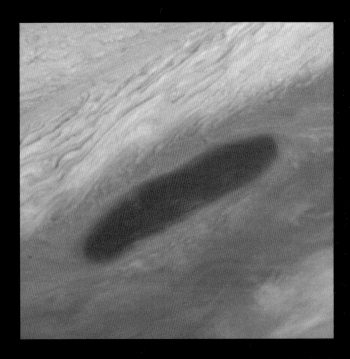

Jupiter's northern pole

11–12 December 2000

While on its way to Saturn, the narrow-angle camera aboard
the Cassini-Huygens space probe took the pictures that together
make up this composite map of the planet's northern pole.
The camera took shots each hour for nine hours while the planet
rotated beneath the spacecraft.

Photo: NASA/JPL/Space Science Institute

⊕

Jupiter's southern pole

11–12 December 2000

Both of these images represent the most detailed colour maps yet
made of Jupiter. The parallel white and reddish-brown circles are
cloud features, while – on this composite image of the planet's southern
pole – the Great Red Spot is very visible in the top left quarter.
What's the Great Red Spot? Read on…

Photo: NASA/JPL/Space Science Institute

The Great Red Spot

This extraordinary reprocessed view of the Great Red Spot was originally made by Voyager 1 on its fly-by in 1979. The spot is actually an ongoing anticyclone storm that measures up to 40,000km across, which scientists estimate may have been ranging for as long as 350 years.

Photo: NASA/Bjorn Jonsson

Jupiter

⊕

Jupiter 'scar' seen in infrared

20 July 2009 (above);
16 August 2009 (below)

This pair of images, taken by NASA's Infrared Telescope Facility on Mauna Kea, Hawaii, show the scattering of particle debris after an object entered Jupiter's atmosphere on 19 July 2009. Appearing as a bright orange spot in the bottom left of the image taken on the day following the impact, nearly a month later the debris – sheared open by Jupiter's winds – has dispersed to form an elongated 'scar'.

Photos: NASA/IRTF/JPL-Caltech/
University of Oxford

⊕

Jupiter in ultraviolet

21 July 1994

The dark spots near the bottom of this ultraviolet image show the
impact on Jupiter's atmosphere caused by fragments of comet
Shoemaker-Levy 9, which spectacularly collided with the planet
over the duration of a week in July 1994. The solitary dark spot
clearly visible just above the centre of Jupiter is nothing to do with
the comet's impact; it's actually Io, one of the planet's many moons.

The moon Io

The Galilean satellites – the four largest of Jupiter's 67 moons – were discovered in 1610 by the Italian astronomer Galileo Galilei. This is Io, the closest of Jupiter's Galilean moons. Home to more than 400 active volcanoes, it is the most volcanically active body in the Solar System.

Photos of all moons: Hubble Space Telescope Comet Team and NASA/ESA

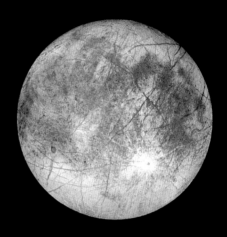

The moon Europa

It might be the smallest of the four Galilean moons, but Europa is still the Solar System's sixth largest moon. It's estimated to be around 4.5 billion years old (around the same age as Jupiter itself) and its surface is covered by a layer of ice, suggesting an active ocean lies beneath.

The moon Ganymede

Larger than both Mercury and Pluto, Ganymede is the largest satellite in the entire Solar System, making it visible to the naked eye. The third Galilean moon from Jupiter, each of its orbits takes just over seven days.

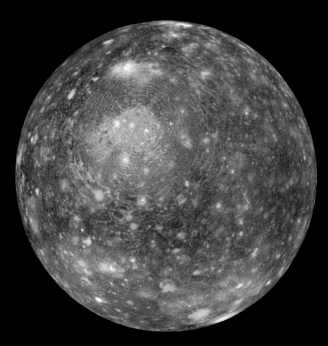

The moon Callisto

Describing Callisto's surface as pockmarked is an understatement; it has more craters than any other object in the Solar System. With little evidence of plate tectonics or volcanic activity, these are believed to be impact craters, the result of being hit by other objects.

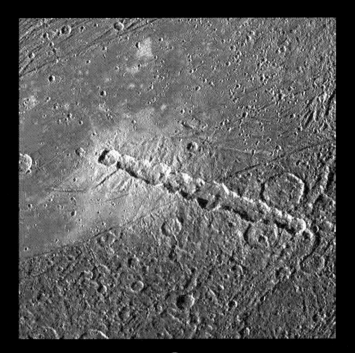

⊕

Crater chain on Ganymede

5 April 1997

This line of 13 craters on the surface of the largest Jovian moon Ganymede is known as Enki Catena and was probably formed by the dispersing parts of a comet impacting in rapid succession.

Photo: Galileo Project/Brown University/JPL/NASA

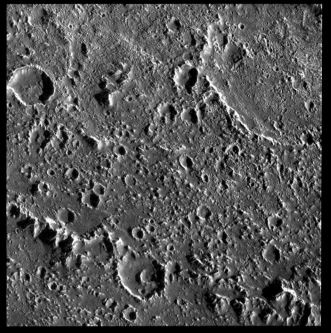

⊕

Callisto's textured terrain

17 September 1997

Here in the Asgard Basin of the Galilean moon Callisto, the contrast between the surface of the impact basin and that of the surrounding plains is clear to see, with small, icy bumps creating a finer texture than elsewhere on the moon.

Photo: NASA/JPL

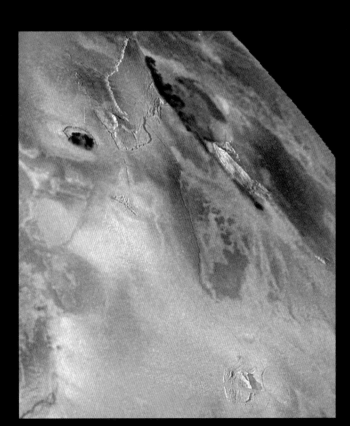

Volcanic activity on Io

3 July 1999 and
25 November 1999

This composite of low-resolution colour and high-resolution black-and-white images taken on two separate dates shows Zal Patera, a large volcanic depression on Io's surface. The darker part indicates the edge of the volcanic crater, while the red area suggests lava erupting onto the moon's surface.

Photo: NASA/JPL/University of Arizona

The surface of Europa

February & December 1997

This mosaic shows the rafts of ice on Europa's surface in its Conomara Chaos region. These blocks have shifted and rotated as the moon's crust has broken up and moved. The background image was taken during the Galileo spacecraft's sixth orbit of Jupiter in February 1997, while the detailed, high-definition images were captured on the probe's twelfth orbit 10 months later.

Photo: NASA/JPL

The launch of Galileo

18 October 1989

The Galileo spacecraft blasts into the Florida skies, carried by Space Shuttle Atlantis. Much of its mission involves studying Jupiter and its moons, where it arrives more than six years later on 7 December 1995. As well as becoming the first spacecraft to orbit Jupiter, Galileo also witnessed the collision of comet Shoemaker-Levy 9 with the planet in 1994.

Photo: NASA

Rings of Saturn

6 October 2004

Saturn takes 29.7 years to complete its orbit of the Sun. During that time, the aspect we see from Earth changes – twice during that period we see the rings edge-on, and for the rest we have clearer views of its northern or southern hemispheres. This high-resolution composite image was pieced together from 126 frames captured by NASA's Cassini craft, and depicts the planet's southern hemisphere and rings.

Photo: Mattias Malmer/Cassini
Imaging Team

Saturn

The sixth planet, Saturn, is also the second largest – at 760 times bigger than Earth, and 9.5 times farther away from the Sun, it is the most distant world that can be seen with the naked eye. Almost entirely hydrogen, it is so light that – if you could find a swimming pool big enough to test the theory – the entire planet would float in water.

Saturn's most famous features are its rings. Almost pure water ice, the nine main rings extend to 80,000km above the surface of the planet that is itself 120,500km in diameter. Surprisingly, they are only 1km thick. The gravity of 62 moons helps to shepherd the ice into defined orbits, leading to the intricate patterning of the rings.

How exactly the rings were created is still a mystery. Some theories suggest that they comprise material left over from the planet's formation, others that they are the shattered remains of a moon. It's uncertain if the rings are even a permanent fixture or merely a passing feature that we are lucky enough to glimpse. **S**

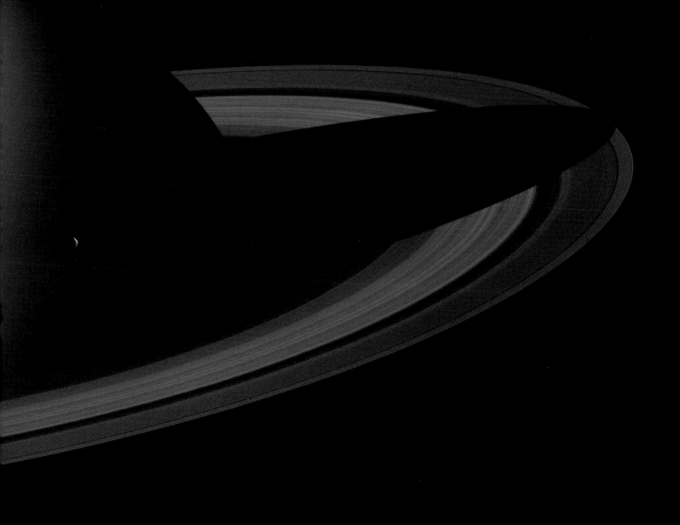

The Rose: Saturn's north polar vortex

27 November 2012

In this false-colour image, in which red indicates lower clouds and green depicts high ones, the vast storm swirling above Saturn's north pole resembles a red rose – albeit one stretching 2,000km across and with winds blasting at over 500km/h. The image was captured by the narrow-angle camera on NASA's Cassini spacecraft – the first time this pole has been photographed.

Photo: NASA/JPL-Caltech/Space Science Institute

Northern storm

11 January 2011

The head of this vast convective thunder-and-lightning storm, first detected by Cassini in December 2010 and depicted in this false-colour image, moved west around the planet over a period of nearly nine months. It formed a vortex up to 12,000km across and eventually consumed itself when it made a full circuit of the planet – about 300,000km – and hit its own tail.

Photo: NASA/JPL-Caltech/SSI/ Hampton University

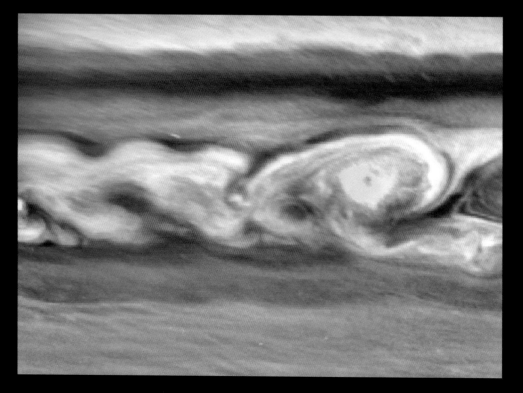

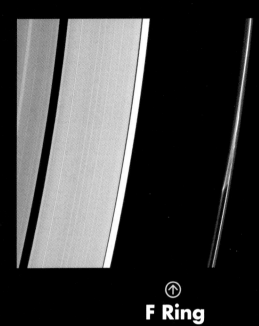

⊕ F Ring

19 October 2013

The outermost discrete ring, about 140,000km from Saturn's centre, is held together by two 'shepherd' moons, Prometheus and Pandora, orbiting either side. This Cassini image shows a strand peeling away from the ring, possibly due to a collision with a small object.

Photo: NASA/JPL-Caltech/Space Science Institute

⊕ Spokes in the B Ring

20 December 2012

Ghostly spokes – subtle lines visible in the darker central area of this Cassini image – appear periodically in the B Ring, usually just after that part has left the planet's shadow. They stretch up to 16,000km long, but last only an hour or two before disappearing.

Photo: NASA/JPL-Caltech/Space Science Institute

⊖ Dark side of the rings

10 November 2012

The bright point visible between the rings in the upper right portion of this image is Venus, showing as just a bright dot at this vast distance. This true-colour picture depicts the unilluminated side of the rings – which are just 10m thick in places – silhouetted against the face of Saturn.

Photo: NASA/JPL-Caltech/Space Science Institute

↑ Titan

26 October 2004

By far the biggest of Saturn's 62 known moons is Titan – 5,150km in diameter and nearly twice the size of our Moon. Titan has a cold, dense atmosphere, and lakes of methane and ethane on its surface.

Photo: NASA/JPL/Space Science Institute

↑ Rhea

21 November 2009

Saturn's second-largest moon is 1,528km in diameter and has a heavily cratered surface. It's thought that Rhea's core may be largely water ice, and some data suggest that it has faint rings.

Photo: NASA/JPL/Space Science Institute

↑ Dione

26 December 2009

This icy moon – at 1,123km across, Saturn's fourth-largest – is known for the wispy-looking features on its trailing hemisphere, which are dramatic chasmata: tectonic fractures or canyons.

Photo: NASA/JPL/Space Science Institute

↑ Tethys

14 October 2009

The surface of Tethys, another water-ice-rich moon measuring 1,062km in diameter, is pitted with craters. The crater called Penelope, 207km wide, is clearly visible on this Cassini image.

Photo: NASA/JPL/Space Science Institute

⊕
Iapetus

10 September 2007

The cratered leading hemisphere of Iapetus – at 1,470km across,
Saturn's third-largest moon – is stained with a coating of dark
organic material. Iapetus also has a pronounced equatorial ridge.

Photo: NASA/JPL/SSI/AstroArts.org

⊕
Enceladus

10 March 2012

Saturn's sixth-largest moon, 504km across, resembles a snowball
– its pale surface is covered with snow and ice, and water vapour
and ice erupt from fractures in its southern regions.

Photo: NASA/JPL-Caltech/Space Science Institute

⊕
Mimas

13 February 2010

The 130km-wide crater named Herschel occupies a sizeable
swathe of the surface of 396km-diameter Mimas; it plunges to
10km deep, and its central peak rises 6km above the crater floor.

Photo: NASA/JPL/Space Science Institute (Cassini)

The day Earth smiled

19 July 2013

Taking advantage of the protection from the Sun's damaging rays afforded by Saturn's shadow, NASA's Cassini craft took a series of images, 141 of which were combined to create this extraordinary mosaic picture. As well as the beautifully defined rings, seven of Saturn's moons are visible, along with Venus, Mars and even Earth, below the rings to the right of the image.

Photo: NASA/JPL-Caltech/SSI

Earth

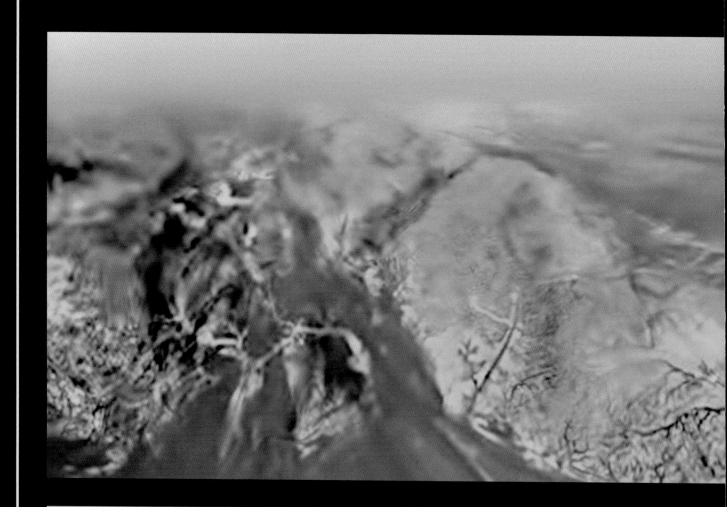

Surface of Titan

14 January 2005

The European Space Agency's Huygens probe, delivered by the Cassini craft, captured the data used to create this Mercator projection of Titan's surface as it descended through the moon's nitrogen-rich atmosphere. These composite images depict a rugged landscape, one eroded by rivers of methane and striated with regions of dark hydrocarbon dunes.

Photo: ESA/NASA/JPL/University of Arizona

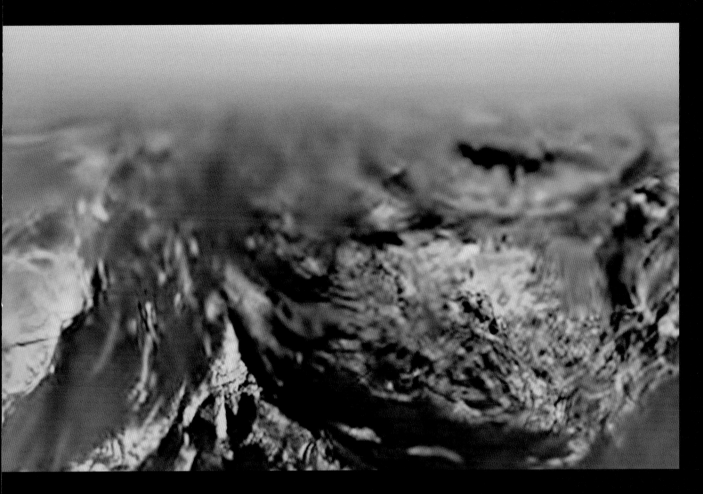

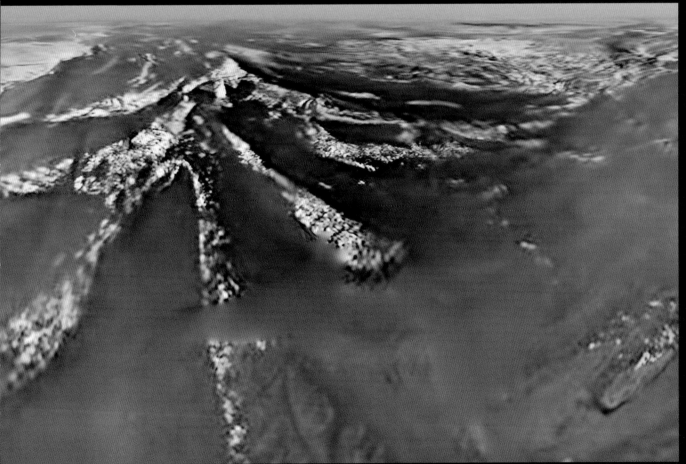

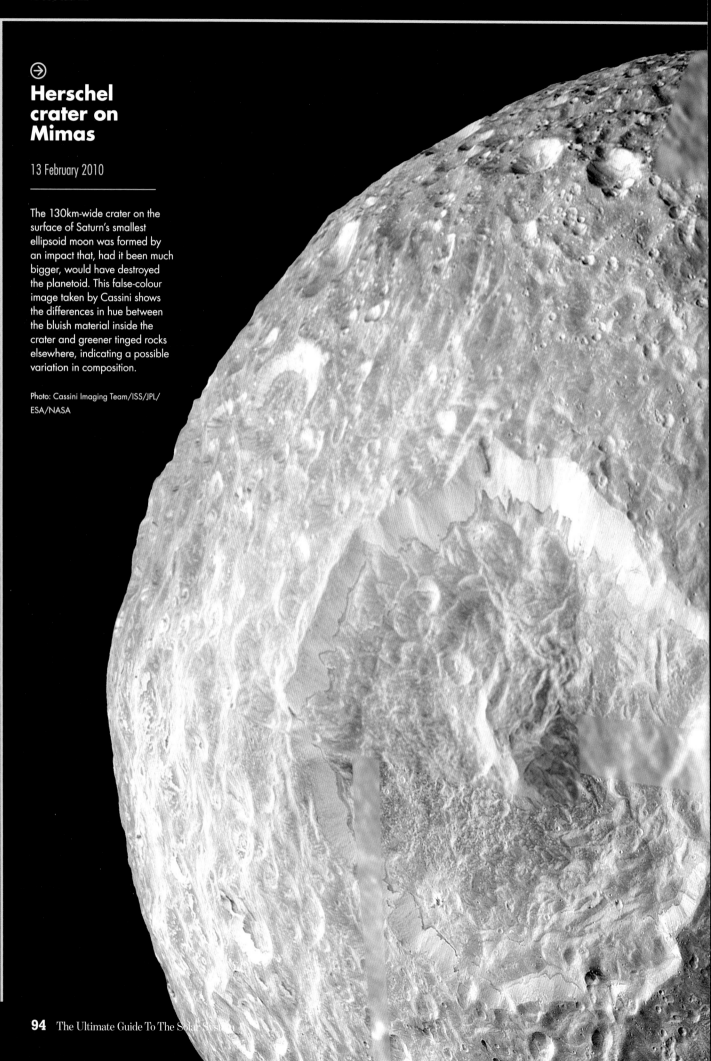

⊕→

Herschel crater on Mimas

13 February 2010

The 130km-wide crater on the surface of Saturn's smallest ellipsoid moon was formed by an impact that, had it been much bigger, would have destroyed the planetoid. This false-colour image taken by Cassini shows the differences in hue between the bluish material inside the crater and greener tinged rocks elsewhere, indicating a possible variation in composition.

Photo: Cassini Imaging Team/ISS/JPL/ ESA/NASA

⊕

Cassini is prepared for launch

22 July 1997

NASA's Cassini craft launched in October 1997, reaching Saturn
in 2004. Its first four-year mission included deploying the European
Space Agency's Huygens probe onto the surface of Titan.

Photo: NASA

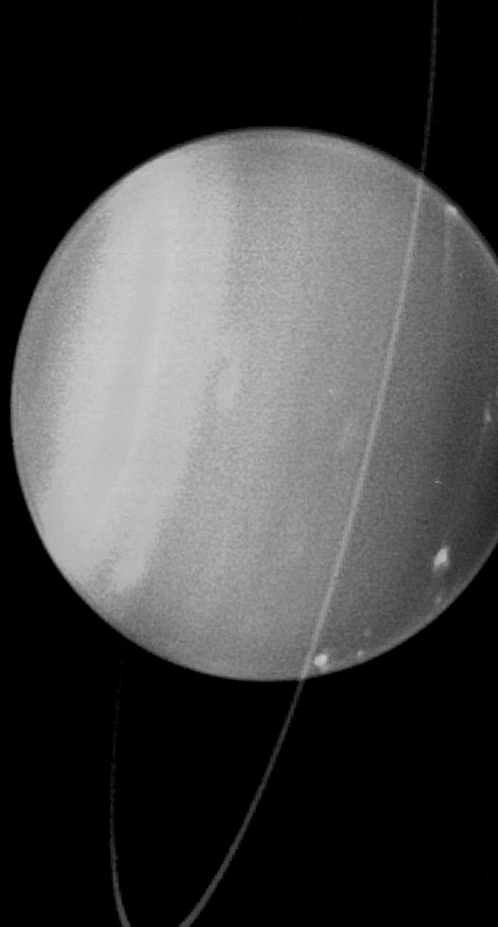

⊕➔

Tilt of
Uranus

11–12 July 2004

This image, gathered by the
10m Keck Telescope in Hawaii,
clearly shows Uranus's rings
and its axial tilt of 97.77°,
possibly caused by a collision
with a large body early in its
existence. Because of this tilt,
and the planet's long Solar
orbit – 84 Earth years – each
hemisphere spends long periods
in darkness. The northern
hemisphere at left is shown
coming into the light; bright
white and blue patches in the
southern hemisphere are clouds
above the atmosphere.

Photo: California Association for
Research in Astronomy/Science
Photo Library

Uranus and Neptune

At the edge of our Solar System lie two distant planets that we call ice giants. They are so cold that their rocky cores have become wrapped in a layer of water ice, hidden beneath a thick hydrogen atmosphere.

The first is Uranus. Unlike the other giant planets, this distant world's clouds seem calm and featureless, coloured pale blue by methane in the upper atmosphere. Uranus is tilted onto one side, meaning that it takes an entire 84-year orbit around the Sun for the whole of the planet to feel sunlight.

The second ice giant, Neptune, is also the farthest planet from the Sun – 4.5 billion kilometres away. But this planet experiences the highest wind speeds in the Solar System, reaching up to 2,400km/h.

These two distant worlds have only ever been visited once, when the Voyager 2 probe passed them in the late 1980s. With no plans for further visits, we will have a long wait to learn more about these remote giants. **S**

⊕

Uranus's atmosphere in detail

25–26 July 2012

The enhanced contrast in this paired picture, taken by the Keck II Telescope in Hawaii, reveals new details including a swarm of convective storms at the north pole (far right) and a distinctive scalloped belt of clouds around the equator.

Photos: Lawrence Sromovsky, Pat Fry, Heidi Hammel, Imke de Pater/University of Wisconsin-Madison

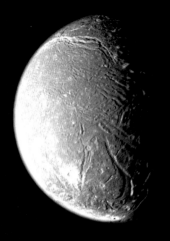

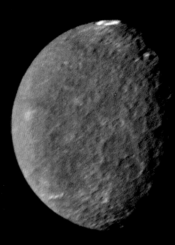

Moons of Uranus

24 January 1986

Voyager 2 captured images of all five of Uranus's ellipsoidal moons – from above left: Ariel, Miranda, Oberon, Titania (the largest) and Umbriel – named after characters from Shakespeare and Pope.

Photos: NASA/JPL

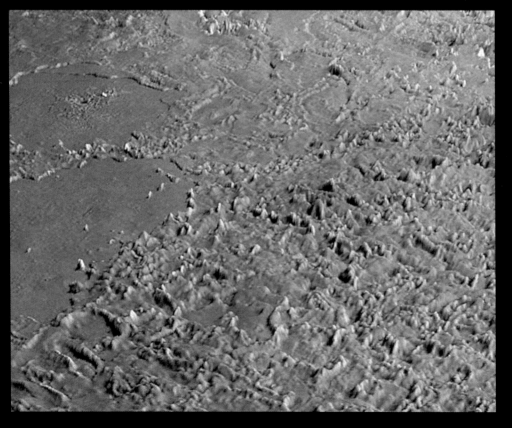

Volcanic plains of Triton

24 August 1989

This view of Triton's surface was produced using topographic data from Voyager 2's flyby, with relief exaggerated by a factor of 25. It shows the rugged landscape that appears as 'canteloupe' terrain in the image opposite; the foreground mounds are diapirs (rising lumps of ice) that erupted from the surface during buckling of the crust. Triton's surface is pocked with faults, volcanic pits and lava flows, and may still be volcanically active.

Photo: NASA/JPL/Universities Space Research Association/Lunar & Planetary Institute

Photo: NASA/JPL/U.S. Geological Survey

Triton

25 August 1989

This mosaic, created from images taken by Voyager 2, has colour synthesised to highlight various surface features, including pinkish deposits on the southern polar cap believed to contain methane ice.

Photo: NASA/JPL/U.S. Geological Survey

Neptune

20 August 1989

Expecting to find a frigid, dormant world, scientists were amazed when images such as this one, captured by the Voyager 2 probe as it flew past before leaving the Solar System, showed Neptune's turbulent atmosphere experienced massive storms and winds reaching 2,400km/h – the strongest observed on any planet or moon.

Photo: NASA

(→)

The Great Dark Spot

25 August 1989

Images transmitted by Voyager 2 showed an anticyclonic storm reaching 13,000km in length, spinning anticlockwise and travelling westward at nearly 1,200km/h. Unlike the equivalent spot on Jupiter, though, this Great Dark Spot is not permanent – the Hubble Space Telescope later revealed that such storms arise and vanish relatively frequently on Neptune.

Photo: NASA/JPL

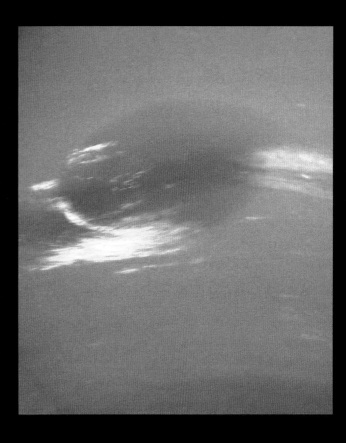

White clouds over Neptune

25–26 June 2011

These images, taken by NASA's Hubble Space Telescope at four-hour intervals over a 16-hour period, cover a full rotation of Neptune. They reveal high-altitude clouds of methane ice crystals, shown as a pinkish hue because they reflect near-infrared light.

Photos: NASA/ESA/Hubble Heritage Team (STScI/AURA)

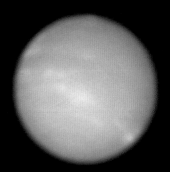

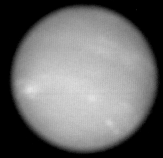

Linear clouds over Neptune

25 August 1989

Voyager 2 captured images of long linear clouds, similar to cirrus formations seen on Earth, stretching for hundreds of kilometres along lines of constant latitude. This image shows vertical relief in the clouds, some of them rising 50km above the base cloud layer.

Photo: NASA/JPL/Planetary Photojournal

Scooter and spots

24 August 1989

The bright feature between the Great Dark Spot and the lower Small Dark Spot is named Scooter, for its rapid movement across the surface; it is believed to be a storm or a patch of cirrus clouds.

Photo: NASA/Voyager 2 Team

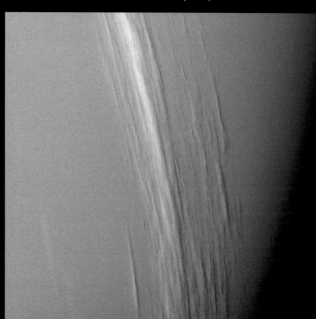
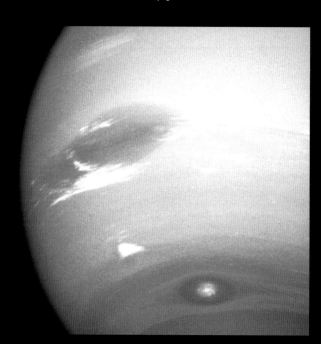

⊕

Voyager

The twin Voyager craft (this illustration depicts Voyager 1) were launched in 1977, with an initial mission to visit Jupiter and Saturn; Voyager 2 went on to fly past Nepune in 1986 and Uranus in 1989, while Voyager 1 entered interstellar space on 25 August 2012. It has travelled farther from Earth than any other human-made object, and will continue to function until 2025, when power from its radioactive generators will dwindle too low to power its instruments.

Photo: NASA

An artist's impression of Quaoar

Discovered in 2002 by Chad Trujillo and Michael Brown from the California Institute of Technology, the dwarf planet Quaoar takes 288 years to orbit the Sun 6.5 billion kilometres away. At the time of its discovery, Quaoar became the largest object found in the Solar System since Pluto was discovered in 1930. The latest estimate put Quaoar's diameter at 1,074km, around a twelfth of that of Earth.

Photo: Science@NASA

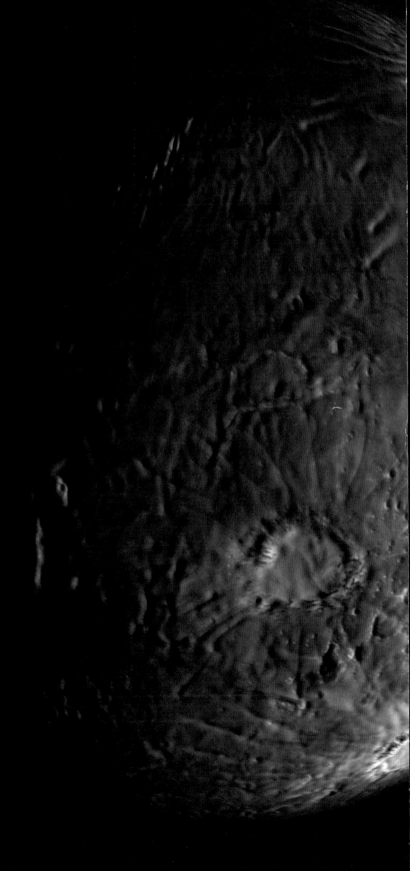

The Kuiper Belt and beyond

The Kuiper Belt surrounds the whole of the Solar System. Starting at Neptune and stretching out to 7.5 billion kilometres from the Sun, this ring of small bodies represents the remnants of the Solar System's formation.

The belt is home to many dwarf planets, the largest of which are Eris and Pluto.

In 2015, the New Horizons mission will reach Pluto, giving us our first direct look at a Kuiper Belt object. However, we have already seen some of the region's former inhabitants up close – most comets originated here and, either through collisions or the tug of the inner planets, were sent into a closer orbit around the Sun.

Kuiper Belt comets all take less than 200 years to orbit the Sun, yet many have much longer orbit times. These must have come from somewhere much more distant, an unseen region called the Oort Cloud. This giant cocoon of objects is thought to extend halfway to the nearest star and marks the very end of our Solar System. Ⓢ

Trans-Neptunian object

This piece of Solar System debris – as rendered by an artist – is an example of a trans-Neptunian object, which range from 40km to 100km across. The star shining brightly in the distance is the Sun, as seen from more than 4.8 billion kilometres away.

Photo: NASA

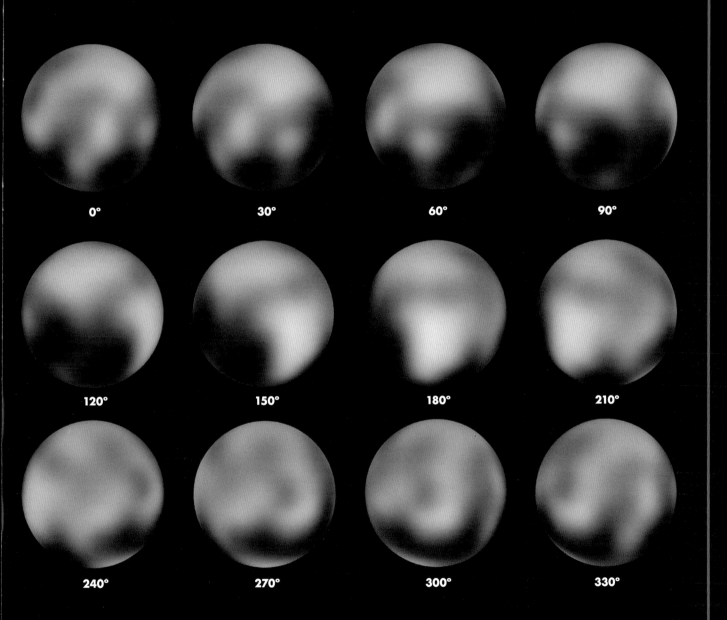

0°   30°   60°   90°

120°   150°   180°   210°

240°   270°   300°   330°

Pluto from all angles
2002–3

Compiled using multiple images taken by the NASA Hubble Space
Telescope between 2002 and 2003, these 12 images reveal the entire
surface of Pluto. Originally declared as the ninth planet on its discovery
in 1930, it was reclassified as a dwarf planet in 2006 having failed
to meet redrawn conditions for planet status.

The interior of Pluto

This cross-section of Pluto shows the core, mantle and crust of the dwarf planet. Observations made using the Hubble Space Telescope suggest that Pluto's interior consists of about 50–70 per cent rock and 30–50 per cent ice. Its temperature varies depending on its position in its orbit, Pluto's surface is thought to be the coldest place in the Solar System at around –225°C.

Photo: Science Source/Science Photo Library

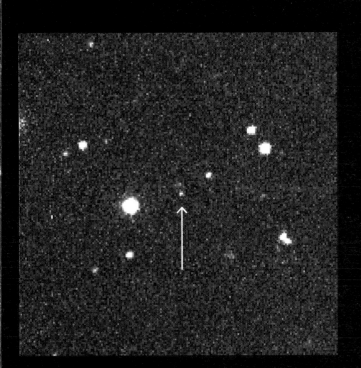

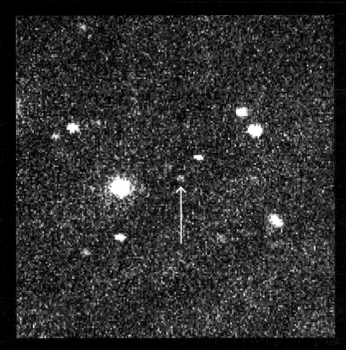

⬆ The first appearance of Sedna

14 November 2003

These three photographs represent the discovery of Sedna. The faint spot indicated by the arrows is the Pluto-like dwarf planet identified by its slight shift in position over these three panels. It is estimated that a single orbit of the Sun takes Sedna around 10,500 years to complete.

Photos: NASA/Caltech

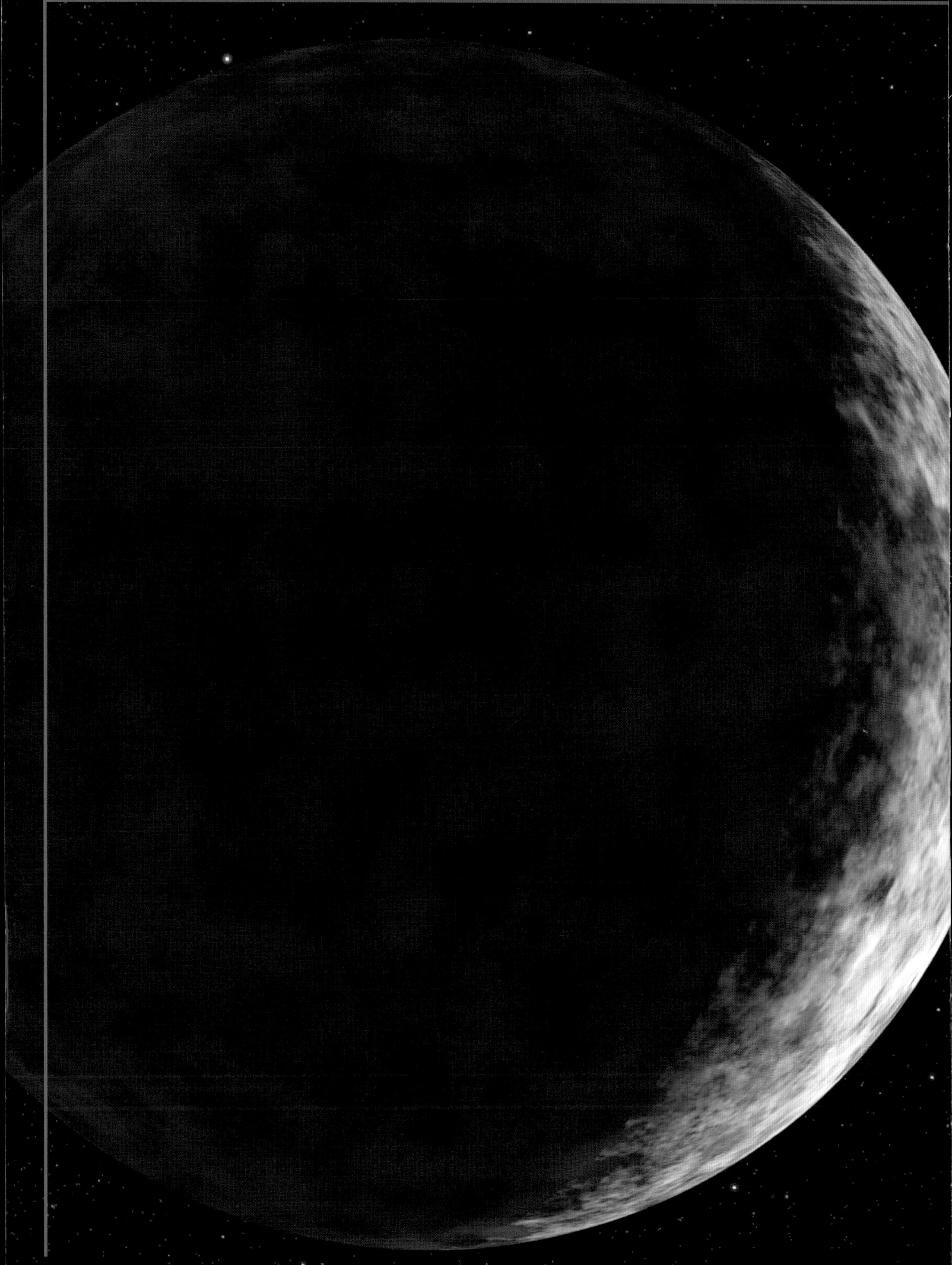

Dwarf planet Eris and its moon Dysnomia

Discovered in 2005, Eris is the Solar System's largest known dwarf planet. More than a quarter larger than Pluto, it was originally regarded as the 10th planet before it – and Pluto – were reclassified a year after its discovery. This artist's impression includes Dysnomia, Eris's only known moon. The distant star is the Sun, which can be up to 16 billion kilometres away during Eris's orbit.

Photo: CalTech/NASA

SKY AT NIGHT MAGAZINE
SUBSCRIPTION ORDER FORM SSBK14

PAYMENT OPTIONS

☑ I would like to subscribe to *BBC Sky at Night Magazine* by Direct Debit and receive my first 5 issues for just £5*

Please complete the form below

Instructions to your Bank or Building Society to pay by Direct Debit	DIRECT Debit

To: The Manager (Bank/Building Society)

Address

Postcode

Name(s) of account holder(s)

Bank/Building Society account number

Branch sort code

Reference number (internal use only)

Originator's identification number

7 1 0 6 4 4

Instructions to your Bank or Building Society. Please pay Immediate Media Company Bristol Ltd Direct Debits from the account detailed in this instruction subject to the safeguards assured by the Direct Debit Guarantee. I understand that this instruction may remain with Immediate Media Company Bristol Ltd and, if so, details will be passed electronically to my Bank/Building Society.

Signature

Date   /   /

Banks and Building Societies may not accept Direct Debit mandates from some types of account

YOUR DETAILS

Title.................First name.....................................

Surname...

Address line 1..

Address line 2..

Town...

Postcode..

Home telephone number.......................................

Mobile telephone number**..................................

Email address**...

Immediate Media Company Limited (publisher of *BBC Sky at Night Magazine* under licence from BBC Worldwide) would love to keep you informed by post or telephone of special offers and promotions from the Immediate Media Company Group. Please tick if you'd prefer not to receive these ☐
** Please enter this information so that *BBC Sky at Night Magazine* may keep you informed of newsletters, special offers and other promotions by email or text message. You may unsubscribe from these at any time.
Please tick here if you'd like to receive details of special offers from BBC Worldwide via email ☐

OTHER PAYMENTS For a year's subscription (12 issues)

☐ UK cheque/credit/debit card – £44.90 for 12 issues **SAVE 25%**

☐ Europe – £69 for 12 issues

☐ Rest of world – £79 for 12 issues

☐ I enclose a cheque made payable to Immediate Media Co Bristol Ltd £...............

CREDIT CARD / DEBIT CARD

Visa ☐ Mastercard ☐ Maestro ☐

Card no. ☐☐☐☐☐ ☐☐☐☐☐ ☐☐☐☐☐ ☐☐☐☐☐

Issue no. (Maestro only) ☐☐ Expiry date ☐☐ ☐☐

Signature................................... Date.......................

*5 issues for £5 only available to UK residents paying by Direct Debit. After your trial period your payments will continue at £20.95 every 6 issues, saving 30% on the shop price. If you cancel within 2 weeks of receiving your fourth issue you will pay no more than £5. Your subscription will start with the next available issue
Offer ends: 5 October 2014

FREEPOST UK ORDERS TO:
BBC Sky at Night Magazine, FREEPOST LON16059, Sittingbourne, Kent, ME9 8DF

POST OVERSEAS ORDERS TO:
BBC Sky at Night Magazine, PO BOX 279, Sittingbourne, Kent, ME9 8DF, UK

SUBSCRIBE TODAY BY DIRECT DEBIT AND...

▶ Try your first **5 issues for just £5***

▶ Continue at just **£20.95 every 6 issues** by Direct Debit, saving 30%

▶ Receive **FREE UK delivery** direct to your door

▶ **Never miss a sell-out issue** of the UK's biggest and best astronomy magazine

DON'T MISS OUT
SUBSCRIBE TODAY!

What lies beyond?

⊘

Bow shock in the Orion Nebula

February 1995

Solar wind from the young star LL Ori formed this bow shock – a three-dimensional wave created as charged particles from the star hit a dense gas cloud. This image, taken by the Hubble Space Telescope, shows an area at the edge of the Great Nebula in Orion – about 1,300 light years away. It offers a clue to what the bow shock from our own Sun's stellar wind might look like when viewed from the vastness of interstellar space.

Photo: NASA/Hubble Heritage Team (STScI/AURA)